Dynamical Chaos

DYNAMICAL CHAOS

Proceedings of
a Royal Society Discussion Meeting
Held on 4 and 5 February 1987

Organized and Edited by
M. V. BERRY, F.R.S., I. C. PERCIVAL, F.R.S.,
AND N. O. WEISS

Princeton University Press
Princeton, New Jersey

First published in *Proceedings of the Royal Society of London*,
series A, volume 413 (no. 1844), pages 1-199

Clothbound edition published in Great Britain for the Society by
Cambridge University Press
ISBN 0-85403-333-5

Paperback edition published in the United States of America by
Princeton University Press at Princeton, New Jersey
ISBN 0-691-02423-5

CONTENTS

Dynamical Chaos

Chairman's introduction

By E. C. Zeeman, F.R.S.

Mathematics Institute, University of Warwick, Coventry CV4 7AL, U.K.

The understanding of chaos and strange attractors is one of the most exciting areas of mathematics today. It is the question of how the asymptotic behaviour of deterministic systems can exhibit unpredictability and apparent chaos, due to sensitive dependence upon initial conditions, and yet at the same time preserve a coherent global structure. The field represents a remarkable confluence of several different strands of thought.

1. Firstly came the influence of differential topology, giving global geometric insight and emphasis on qualitative properties. By qualitative properties I mean invariants under differentiable changes of coordinates, as opposed to quantitative properties which are invariant only under linear changes of coordinates. To give an example of this influence, I recall a year-long symposium at Warwick in 1979/80, which involved sustained interaction between pure mathematicians and experimentalists, and one of the most striking consequences of that interaction was a transformation in the way that experimentalists now present their data. It is generally in a much more translucent form: instead of merely plotting a frequency spectrum and calling the incomprehensible part 'noise', they began to draw computer pictures of underlying three-dimensional strange attractors.

2. Secondly, classical differential equations remain as important as ever. Although the general theory has seen major advances in the last two decades, every now and then it runs into a brick wall. For example after Thom's spectacular success in the 1960's in proving the density of stable functions and classifying elementary catastrophes, a similar programme was attempted for dynamical systems. Structurally stable systems, however, turned out to be neither dense nor classifiable. Attention has consequently now switched back to examining classical examples with the advantage of new insight. Meanwhile there is a vast mountain of unsolved problems for the mathematician to work on. Even the notion of strange attractors is not, as yet, satisfactorily defined, and will remain so until there is enough theory built upon it to give an appropriate definition due weight.

3. Thirdly, comes the influence of the ergodic theoretical approach, bringing ideas of entropy and averaging to bear upon differential equations. Previously these two fields were studied separately: a system was thought to be either deterministic or ergodic. But now we are familiar with many examples in which the regions of predictability and chaos are closely interwoven, and the transitions between the two are of paramount importance.

4. Fourthly was the advent of fast interactive graphics in computers, which enabled one to perceive patterns within complex systems that might otherwise never have been suspected to be there, and to formulate conjectures that can then be proved by traditional methods. A beautiful example of this has been the use of

[3]

renormalization techniques to study the breakdown of invariant tori and the onset of chaos.

5. Fifthly is the development of precision experiments in chaotic physical systems. For example there have been major advances in the depth of understanding of the onset of turbulence. And the confidence that this has given to physicists and astronomers has opened their eyes to many examples, now well documented, of deterministic chaos amongst natural phenomena.

6. Sixthly, and finally, has come a new understanding of chaotic biological systems. Mathematical modelling in biology tends to be either very simple or very sophisticated. For most research in biology one needs little more mathematical equipment than the integers, but at the other end of the spectrum one needs very sophisticated mathematics, because biology is in principle far more complicated than physics. We shall see in the next few decades a new generation of mathematical biologists beginning to tackle problems in which the complexity is fundamental.

Summarizing: it is the confluence of these ideas makes the subject so rich, and promises us a fascinating meeting.

Diagnosis of dynamical systems with fluctuating parameters

By D. Ruelle

Institut des Hautes Études Scientifiques, 35 Route de Chartres,
91440 Bures-sur-Yvette, France

Many time evolutions occurring in Nature may be considered as non-autonomous, but dependent on parameters that vary slowly with time. It is argued here that some, but not all, of the tools used to understand chaotic dynamics remain useful in this situation.

In recent years it has been ascertained that many time evolutions observed in Nature exhibit the features of *chaos*. This means that they are deterministic time evolutions involving only a finite number of degrees of freedom, but that a complicated non-periodic behaviour is observed, due to *sensitive dependence on initial condition*. Mathematically, the deterministic time evolution corresponds to an autonomous differentiable dynamical system, sensitive dependence on initial condition means that a small perturbation of the initial condition will grow exponentially with time (as long as it does not become too large). The asymptotic evolution of the system takes place on a (usually) complicated set in phase space called a *strange attractor*.

A fundamental finding is that hydrodynamic turbulence is chaotic, and described by strange attractors. Many other examples of chaos have been demonstrated clearly in various areas of physics and chemistry, and less clearly in biology and economics. Investigations of experimental chaos have been mostly of either geometric or ergodic nature. (The study of chaotic power spectra, broad band spectra, has also played an important role historically, but does not at this time yield the sort of detailed information that is provided by other techniques.) The geometric approach visualizes reconstructions of attractors and of bifurcations and is limited to weakly excited systems ('onset of turbulence'). The ergodic approach determines information dimensions, characteristic (i.e. Liapunov) exponents and entropies, and is applicable to moderately excited systems.

The analysis of relatively modest data can provide a usable power spectrum, or an estimate of the information dimension by the Grassberger–Procaccia algorithm†. In general, however, the detailed diagnosis of chaotic dynamical systems requires long time series of high quality (stability of parameters of the system and precision of experimental measurements). One faces then the problem that the systems for which our techniques work best are not those in which we are mostly interested. Among the latter we may quote pulsating variable stars,

---

† See Grassberger & Prococcia (1983); see also Eckmann & Ruelle (1985) for a general review of the ergodic approach.

electroencephalograms and time series of economics. One can of course dismiss at least the last two examples by arguing (reasonably) that the electrical activity of the brain, and the stock market, are not autonomous dynamical systems with few degrees of freedom.

Against this reasonable view let me remark that information dimension estimates for EEGs (electroencephalograms) (see Layne *et al.* 1986; Rapp *et al.* 1987) and time series of economics (see Scheinkman & Le Baron 1986) are not at all suggestive of pure randomness. Let me talk of EEG data that I have seen (some from Rapp and some from Lehmann, analysed in Geneva in collaboration with J.-P. Eckmann and S. Kamphorst). They suggest that there are many degrees of freedom, or 'modes', with decreasing amplitudes, and that computations of information dimension yield variable results depending on which modes have amplitudes sufficiently large to be captured by a given calculation. This type of 'explanation' is, however, basically unsatisfactory because 'modes' cannot in general be separated in a truly nonlinear theory.

What then? I suggest that some interesting time evolutions occurring in Nature, those with *adiabatically fluctuating parameters* (AFPs), although not represented by an autonomous dynamical system, are accessible to analysis. I have in mind evolutions of the type

$$\mathrm{d}x/\mathrm{d}t = F(x, \lambda(t)) \quad \text{(continuous time)}$$

or

$$x_{n+1} = f(x, \lambda(n)) \quad \text{(discrete time)},$$

where the time dependence of λ is assumed to be adiabatic (slow compared with the characteristic times of the autonomous systems obtained by fixing λ), and not too large. Then, instead of a fixed attractor A, we have a family (A_λ) depending on $\lambda = \lambda(t)$ or $\lambda(n)$. The evolution of λ might itself be determined by a dynamical system, but we consider it as arbitrarily given *a priori*. Note that a time evolution of the above type is expected both for EEGs and in economics. Note also that *noise* can be accommodated in our λ-dependence provided that it satisfies the requirement of adiabaticity.

A first remark is that in a system with AFPs the information dimension will be considerably messed up, because instead of looking at an attractor A we are looking at a union $\cup A_\lambda$. The observed information dimension will thus be the dimension of the attractors A_λ (supposed to be independent of λ) plus the dimension of the set of λs in parameter space. If we observe, for instance, an attracting periodic orbit with slowly decreasing amplitude (mechanical oscillations with friction) we shall obtain a dimension equal to $1+1 = 2$. Long-term evolution in economics would similarly increase the dimension by 1. All we can say in general is that the observed dimension is an upper bound to the dimensions of the A_λ (more precisely one should speak of the information dimension of invariant probability measures carried by A_λ). If one suspects (as in economics) that there is a long time 'secular' evolution of the system, this can be checked by taking an *early* point $X(t_0)$ on the reconstruced attractor and looking at the statistics of times at which the point $X(t)$ comes back close to $X(t_0)$. (For a long time series these times will be predominantly at the beginning of the series if there is a secular evolution of the system.)

Contrary to the dimension, the higher characteristic exponents may not be much perturbed by the fluctuation of the parameters λ. In other words, it will often be the case that the higher characteristic exponents are stable *practically* under small changes of λ; note however that this is not a mathematical statement of continuity. (If the time evolution of $\lambda(t)$ is given by a dynamical system, adiabaticity will correspond to small exponents for the λ-evolution, and those will not interfere with the higher characteristic exponents of the global system.) Therefore the determination of the higher characteristic exponent (or exponents) is very desirable for systems with AFPs, because it (or they) can provide more unambiguous information than the information dimension.[†]

We turn now to the problem of short-term predictions for the time evolution of dynamical systems. Consider for simplicity a time series (u_i) corresponding to a system with discrete time, and chose an embedding dimension n such that the points $(x_i, \ldots, x_{i+n-1}) \in \mathbb{R}^n$ give a faithful representation of the dynamical system on its attractor. Then one can determine a continuous function Φ such that

$$u_{i+n} = \Phi(u_i, u_{i+1}, \ldots, u_{i+n-1})$$

(see Ruelle 1987, §3). The introduction of AFPs will not make this representation useless. Therefore short-term predictions of the evolution of a dynamical system with fluctuating parameters remain possible.

References

Eckmann, J.-P., Oliffson-Kamphorst, S., Ruelle, D. & Ciliberto, S. 1987 Liapunov exponents from time series. *Phys. Rev.* A **34**, 4971–4979 (1986).
Eckmann, J.-P. & Ruelle, D. 1985 Ergodic theory of chaos and strange attractors. *Rev. Mod. Phys.* **57**, 617–656.
Grassberger, P. & Procaccia, I. 1983 Measuring the strangeness of strange attractors. *Physica* D **9**, 189–208.
Layne, S. P., Mayer-Kress, G. & Holzfuss, J. 1986 Problems associated with dimensional analysis of electroencephalogram data. In pp. 246–256. *Dimensions and entropies in chaotic systems* (ed. G. Mayer-Kress), pp. 246–256. Berlin: Springer.
Rapp, P. E., Zimmerman, I. D., Albano, A. M., Deguzman, G. C., Greenbaum, N. N. & Bashore, T. R. 1987 Experimental studies of chaotic neural behavior: cellular activity and electro-encephalographic signs. In *Nonlinear oscillations in chemistry and biology* (ed. H. G. Othmer). New York: Springer. (In the press.)
Ruelle, D. 1987 Theory and experiment in the ergodic study of chaos and strange attractors. In *8th Int. Congress on Mathematical Physics* (ed. M. Mebkhout & R. Senior), pp. 273–282. Singapore: World Scientific.
Scheinkman, J. A. & Le Baron, B. 1987 Nonlinear dynamics and stock returns. Chicago preprint.

Discussion

N. O. Weiss (*Department of Applied Mathematics and Theoretical Physics, University of Cambridge, U.K.*). Surely there will be difficulties if one is dealing with a physical system involving several disparate timescales. One might attempt to describe its behaviour by techniques involving separation of scales or averaging, but if one simply increases the embedding dimension there may be spurious results.

[†] See Eckmann *et al.* (1987) for the description of an algorithm for the determination of the characteristic exponents.

For example, solar magnetic activity shows several different timescales, all of which appear to involve chaos. The day-to-day variation (which has been studied by Spiegel and his colleagues) can be separated from the 11-year solar cycle (which is aperiodic) and from the long-term irregular modulation associated with grand minima. No doubt there are other timescales too and each of them could have its own low-dimensional attractor.

D. RUELLE. The choice of a unit of time between measurements, and of the total recording time, operate a certain choice of timescale. The effect of smaller scales may appear as *noise*, and that of longer scales as *drift*. The ideas of this paper would apply to this drifting situation.

Nonlinear dynamics, chaos and complex cardiac arrhythmias

By L. Glass[1], A. L. Goldberger[2], M. Courtemanche[1]
and A. Shrier[1]

[1] Department of Physiology, McGill University, Montreal,
Quebec, Canada H3G 1Y6

[2] Cardiovascular Division, Beth Israel Hospital, Boston, Massachusetts 02215,
U.S.A.

Periodic stimulation of a nonlinear cardiac oscillator *in vitro* gives rise
to complex dynamics that is well described by one-dimensional finite
difference equations. As stimulation parameters are varied, a large
number of different phase locked and chaotic rhythms is observed. Similar rhythms can be observed in the intact human heart when there is
interaction between two pacemaker sites. Simplified models are analysed,
which show some correspondence to clinical observations.

1. Introduction

The normal adult human heart at rest usually beats at a rate of between 50
and 100 times per minute. In many circumstances, some of which are life-
threatening, but most of which are not, the normal rhythmicity is altered, resulting
in abnormal rhythms called cardiac arrhythmias. The point of this paper is to
show that a branch of mathematics called nonlinear dynamics may be useful in the
analysis of physiological processes believed to underlie normal heart rate regulation and some cardiac arrhythmias.

The idea that mathematical analysis can play a role in understanding cardiac
arrhythmias is not novel. Indeed, in the 1920s it was demonstrated that as
parameters in mathematical models for the heart were varied, several different
rhythms that resembled clinically observed arrhythmias could be generated
(Mobitz 1924; van de Pol & van der Mark 1928). In nonlinear mathematics,
these changes in the qualitative features of the rhythms that are observed as parameters vary are called bifurcations. Thus the problem of understanding cardiac
arrhythmias in the human heart is identified with understanding the bifurcations and complex dynamics in mathematical models of the human heart.

One type of dynamic behaviour that is the object of intensive analysis in
mathematics is chaos. Loosely, chaos is defined as aperiodic dynamics in deterministic systems in which there is sensitive dependence to the initial conditions.
This means that although in principle one could determine precisely the future
evolution of the system starting from some initial condition, for chaotic dynamics
any difference in the initial condition, no matter how small, will eventually lead
to marked differences in the future evolution of the system. Although the existence
of chaos was known to Poincaré and others since the end of the last century, in the

past decade there has been a recognition of the potential significance of chaos in understanding the genesis of aperiodic dynamics experimentally observed in the natural sciences (Cvitanovic 1984). Unfortunately, there is in our view not yet an adequate operational definition for chaos in experimental or naturally occurring systems, but see Mayer-Kress (1986) for recent advances. The concept of chaos excludes non-deterministic stochastic processes, such as the Poisson process or random walk. It is not yet known how to measure the relative contribution of chaos as opposed to non-deterministic stochastic processes in experimental data.

Normal individuals show marked fluctuations in heart rate (Kitney & Rompelman 1980; Kobayashi & Musha 1982; Pomeranz et al. 1985; De Boer et al. 1985). In addition, cardiac arrhythmias are often extremely irregular and unstable (Pick & Langendorf 1979; Schamroth 1980). The adjective 'chaotic' is sometimes used to characterize cardiac arrhythmias that are believed to arise when there are several pacemaker sites competing for control of the myocardium (Katz 1946; Phillips et al. 1969; Chung 1977). It has been proposed that chaotic dynamics, in the mathematical sense, may underlie normal heart-rate variability (Goldberger et al. 1984; Goldberger & West 1987) as well as certain cardiac arrhythmias in humans (Guevara & Glass 1982; Smith & Cohen 1984; Glass et al. 1986b). The absence of a clear definition for chaos in experimental data has led to controversy. For example, ventricular fibrillation, an arrhythmia that leads to rapid death, is frequently called chaotic by clinicians, and it has been proposed that it may be associated with chaos in deterministic systems (Smith & Cohen 1984). However, there are marked periodicities during ventricular fibrillation, and the presence of deterministic chaos in this arrhythmia has been questioned (Goldberger et al. 1985, 1986).

In humans it is frequently difficult to analyse the mechanism underlying an arrhythmia, and systematic experimental studies are usually not feasible. One means of analysis is from the electrocardiogram (ECG), a record of electrical potential differences on the surface of the body that reflects the electrical activity associated with the heartbeat. Because the ECG can be obtained with lightweight monitors, it can be readily recorded over long time intervals. The ambulatory (Holter) ECG is an important means for evaluating patients. Holter recordings for as long as 24 h can be readily obtained, but conventional analysis of such records is limited. The great wealth of data about the dynamics of the heart that is contained in such records is generally distilled to characterize the mean heart rate and range. The presence and frequency of abnormal electrocardiographic complexes, which reflect abnormalities in cardiac impulse formation and propagation, are also determined. However, the analysis of long-term fluctuations in the Holter ECG is largely ignored.

One class of arrhythmias that has recently been the subject of much attention results from the presence of two pacemakers: the normal (sinus) pacemaker and a pacemaker at an ectopic (non-sinus) location. Such rhythms, whose existence has been recognized since the start of this century (Fleming 1912; Kaufmann & Rothberger 1917) are now called parasystolic rhythms. The possibility for interactions between the sinus rhythm and the ectopic rhythm often complicates interpretation of such rhythms. However, recent workers have made great progress in developing both experimental (Jalife & Moe 1976; Jalife & Michaels 1985) and theoretical (Moe *et al.* 1977; Swenne *et al.* 1981; Ikeda *et al.* 1983) models for parasystole. Interpretation of ECG records has led to the recognition of the importance of parasystolic mechanisms (Jalife *et al.* 1982; Nau *et al.* 1982; Castellanos *et al.* 1984).

Here we consider the interaction between a fixed periodic stimulus and a cardiac oscillator. Such a problem is of interest because it is amenable to experimental and theoretical analysis and because of its relevance to the interpretation of parasystolic rhythms. In §2 we consider the effects of periodic stimulation of spontaneously beating aggregates of cells from embryonic chick heart (Guevara *et al.* 1981; Glass *et al.* 1983, 1984, 1986b). Theoretical analysis of this system shows that periodic dynamics are expected at some stimulation frequencies and amplitudes, whereas chaotic dynamics are expected for other stimulation parameters. Experiments are in close agreement with the theory. In §3 we develop a theoretical model for parasystole. The model extends previous theoretical models of parasystole (Moe *et al.* 1977; Swenne *et al.* 1981; Ikeda *et al.* 1983; Glass *et al.* 1986a). We describe the bifurcations in the theoretical model and show that chaotic dynamics is expected over some regions of parameter space. In §4 we discuss Holter ECG records from ambulatory patients who display frequent ectopic beats. These records may show extremely irregular dynamics which we discuss in the context of chaotic dynamics and modulated parasystole. Finally, the significance of this approach to the analysis of cardiac dynamics is discussed.

2. PERIODIC STIMULATION OF A CARDIAC OSCILLATOR

In this section we describe the effects of single and periodic stimulation of an aggregate of spontaneously beating cells from embryonic chick heart. As this work has been described in several recent publications, we briefly summarize the main results and refer the reader elsewhere for more details (Guevara *et al.* 1981; Glass *et al.* 1983; Glass *et al.* 1984; Glass *et al.* 1986b; Guevara *et al.* 1986).

Spontaneously beating aggregates of ventricular heart cells are formed by dissociating the ventricles of seven-day embryonic chicks and allowing the cells to reaggregate in tissue culture medium. The resulting aggregates are approximately 100–200 μm in diameter and each beats with its own intrinsic frequency, which lies in a range of about 60–120 times per minute (DeHaan & Fozzard 1975). A glass microelectrode is inserted intracellularly and can be used to inject single and periodic current pulses into the aggregate. In the present context, the electrical stimulator is analogous to the sinus rhythm, and the aggregate is analogous to an ectopic focus. Clearly, this represents a gross oversimplification of the anatomically complex heart, as it in no sense takes into account the spatial heterogeneity of cardiac tissue nor the various feedback mechanisms that act to modulate cardiac activity *in vivo*. Nevertheless, as stimulation parameters are varied, this model system generates a great variety of rhythms that resemble clinically observed arrhythmias. Some of these rhythms are periodic with N cycles of the periodic stimulation for each M cycles of the cardiac oscillation ($N:M$ phase locking).

Other rhythms are aperiodic (figure 1). The dynamics of this system can only be understood by using techniques in nonlinear dynamics. Thus, this model system is useful to fix ideas and to form a foundation for the analysis of more complex situations.

In response to a single pulse of electrical current, the phase of the oscillation is usually reset. The magnitude of the resetting is proportional to the amplitude and the phase of the current pulse. Generally within a few cycles, the rhythm is re-established at the same frequency as before but with a permanent shift of phase. The re-establishment of the same amplitude and frequency of the oscillation following a perturbation, indicates that from a mathematical point of view it should be useful to think of the cardiac oscillation as a stable limit cycle oscillation. A stable limit cycle oscillation represents a periodic solution of a differential equation that is attracting in the limit $t \to \infty$, for points in the neighbourhood of the cycle.

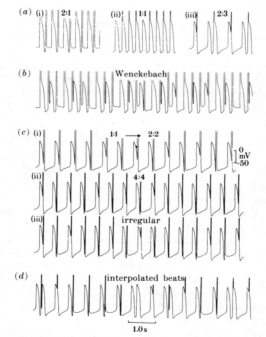

FIGURE 1. Representative transmembrane recordings showing the effects of intracellular periodic stimulation *in vitro* of spontaneously beating embryonic heart cells from chick. The stimulus artifact is observed as a narrow upward deflection. The broader complex is the action potential which corresponds to the contraction of the aggregate. (*a*) Stable phase-locked rhythms; (*b*) rhythms in which the time from the stimulus artifact to the action potential progressively increases until a beat is dropped; this is analogous to the Wenckebach phenomenon in electrocardiology (Pick & Langendorf 1979); (*c*) period-doubling bifurcations and irregular chaotic dynamics; (*d*) irregular rhythm in which there are more action potentials than stimuli. From Guevara *et al.* (1981).

Theoretical analysis of this system is possible by assuming that following a stimulus, the return to the cycle is extremely rapid (figure 2). Thus, if a periodic train of stimuli is delivered to the system with a time interval of T between the stimuli, then the effects of periodic stimulation can be computed from the finite difference equation

$$\phi_{i+1} = g(\phi_i) + \tau \,(\text{mod } 1), \tag{1}$$

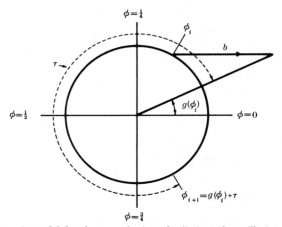

FIGURE 2. A schematic model for the perturbation of a limit cycle oscillation by a periodic stimulus. Provided that the relaxation to the limit cycle following a stimulus is rapid, (1) can be derived.

where ϕ_i is the phase of the ith stimulus and $\tau = T/T_0$, where T_0 is the control cycle length of the aggregate. The function g, called the phase transition curve, depends on the strength of the electrical current and can be measured from the phase resetting resulting from a single stimulus (Perkel *et al.* 1964; Pavlidis 1973; Guevara *et al.* 1981; Glass *et al.* 1983, 1984).

Equation (1) is a finite difference equation and the analysis of bifurcations of such equations is a topic of much current interest. In the present case, the finite difference equation takes a point on the circumference of a circle, ϕ_i, and generates a new point also on the circumference of a circle, ϕ_{i+1} (it is called a circle map). The analysis of circle maps was initiated by Poincaré and major advances in analysing the bifurcations of circle maps were made by Arnol'd (1965) for the case of invertible (for each ϕ_i there is a unique ϕ_{i+1} and vice versa) circle maps. In the practical situations that arise in the experimental system the circle maps are not always invertible and an extension of the theory of invertible circle maps was carried out (Guevara & Glass 1982; Glass *et al.* 1983, 1984; Keener & Glass 1984; Belair & Glass 1985). The analysis of bifurcations of noninvertible circle maps provides a fertile field for mathematical research (for a recent study and references to other work see MacKay & Tresser 1986).

From (1) it is possible to compute the effects of periodic stimulation at any frequency once g, which is measured experimentally, is determined (figure 3). The following are the main conclusions derived from the experimental and theoretical studies. (i) There is a well-defined ordering of phase-locked rhythms corresponding to theoretical predictions based on the analysis of circle maps; (ii) for some stimulation parameters for which one theoretically computes that there should be

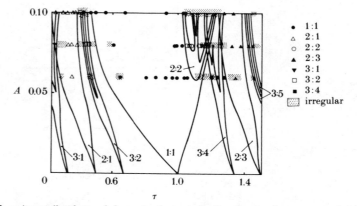

FIGURE 3. Experimentally observed dynamics for periodically stimulated aggregates of chick heart cells superimposed on theoretically computed phase locking zones. The computations use (1) and experimentally measured phase transition curves as described in Glass *et al.* (1984). τ represents the period of the stimuli divided by the period of the oscillations in the aggregates. A is the amplitude of the stimulus in arbitrary units. The circle map in (1) is invertible for $0 < A < 0.039$ and the Arnol'd tongue structure is observed. From Glass *et al.* (1984).

chaotic dynamics, aperiodic dynamics are experimentally observed; (iii) for situations in which the dynamics are believed to be chaotic, if ϕ_{i+1} is plotted as a function of ϕ_i from experimental data then the results are in good agreement with maps calculated based on single pulse phase resetting studies. Thus, our ability to compute theoretically the bifurcations for this system, and the strong agreement between theory and experiment, gives us confidence that the aperiodic dynamics in some regions of parameter space would still be present even if it were possible to eliminate all environmental noise (i.e. the dynamics is chaotic for some parameter values).

3. THEORETICAL MODELS FOR PARASYSTOLE

In parasystole there is competition between the normal sinus pacemaker and a pacemaker which is present at some ectopic (i.e. non-sinus) focus. Although the ectopic focus can be present in either the atria or ventricles, for the current discussion we assume that the ectopic focus is present in the ventricles. The recognition of the possibility of ventricular parasystole dates back at least as far as Fleming (1912) who based his work on the analysis of pulse pressure data. In the ideal situation the two rhythms have their own set frequencies and there is no phase resetting of the ectopic focus by the sinus rhythm. This 'pure' parasystole has recently been analysed (Glass *et al.* 1986*a*) and we follow the treatment there. It is also possible that the sinus rhythm can act to modulate the ectopic rhythm (Jalife & Moe 1976; Moe *et al.* 1977; Swenne *et al.* 1981; Ikeda *et al.* 1983). For this case of 'modulated 'parasystole we follow the basic ideas sketched out in these earlier papers, but try to place the analysis in the context of current studies in nonlinear dynamics and give some new computations. The above formulations assume that parameters remain constant. In realistic situations, the parameters may in fact fluctuate. Accordingly, we consider some effects of parameter fluctuation in the above models.

(a) Pure parasystole

We assume the mechanism for parasystole considered by Fleming (1912) and Kaufman & Rothberger (1917); figure 4. There is a normal sinus rhythm with period t_s and an ectopic rhythm with a period t_e, where $t_e > t_s$. After each sinus beat there is a refractory period θ. If the ectopic focus generates an impulse during the refractory period it is blocked, but otherwise it will lead to an ectopic beat which can be recognized on the electrocardiogram because of its abnormal morphology. After each ectopic beat, the next sinus beat is assumed to be blocked, resulting in a 'compensatory pause'.

FIGURE 4. Schematic model for pure parasystole. Sinus rhythm (s) and ectopic rhythm (e) are shown. Refractory time is represented as a shaded region. Any ectopic beat that falls outside the refractory time is conducted (filled arrows) and leads to a blocking of the subsequent sinus beat (dashed lines). Ectopic beats falling during the refractory time are blocked (open arrows). In the illustration $\theta/t_s = 0.4$, $t_e/t_s = 1.65$, and there are either 1, 2 or 4 sinus beats between ectopic beats. From Glass *et al.* (1986*a*).

Remarkably, the hypothesized mechanism for pure parasystole is equivalent to a well-studied problem in number theory (Slater 1967) and a very detailed analysis of the dynamics for fixed t_e, t_s and θ can be given (Glass *et al.* 1986*a*). In particular, we have found the following rules for parasystole.

Rule 1. For any ratio of t_e/t_s there are at most three different values for the number of sinus beats between ectopic beats.

Rule 2. One and only one of these values is odd.

Rule 3. For any value of t_e/t_s at which there are three different values for the number of sinus beats between ectopic beats, the sum of the two smaller values is one less than the larger value.

Rule 4. Consider the sequence giving the number of sinus beats between ectopic beats. One and only one of these values can succeed itself.

To illustrate these rules we have numerically computed the sequences giving the number of sinus beats between ectopic beats for fixed parameter values. For any fixed set of parameters call $p(a)$, $p(b)$ and $p(c)$ the probability that there are a, b

or c sinus beats between ectopic beats, where $p(a)+p(b)+p(c) = 1$. In figure 5 we display these probabilities as a function of t_e/t_s for $\theta/t_s = 0.4$. In figure 6 we show the number of sinus beats between ectopic events in the $(t_e/t_s, \theta/t_s)$ plane. The regions that are not labelled contain smaller zones which can be readily determined by using the procedures in Glass *et al.* (1986*a*). Some of the features theoretically

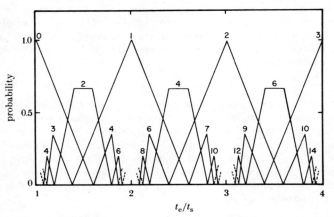

FIGURE 5. Histograms showing relative numbers of sinus beats between ectopic beats for pure parasystole for $\theta/t_s = 0.4$. From Glass *et al.* (1986*a*).

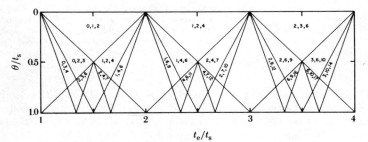

FIGURE 6. Allowed values for the number of sinus beats between ectopic beats for pure parasystole. Allowed values in the unlabelled regions can be determined from the construction described in Glass *et al.* (1986*a*). From Glass *et al.* (1986*a*).

predicted can be found in published reports of parasystolic rhythms. For example Kinoshita (1978, case 7) and Schamroth (1980, case 79) report patients who display either 1, 2 or 4 sinus beats between ectopic beats for parameters that fall in the 1, 2, 4 zones in figure 6 and Lightfoot (1978) describes transitions that arise as the sinus frequency varies that are also consistent with this figure. However, these reports as well as others in the literature, are not consistent with all the four rules above. Deviations from the rules of pure parasystole would be expected if there was modulation of the ectopic rhythm by the sinus beat, and also if there were fluctuations in the sinus or ectopic rhythms. We now consider the effects of these modifications.

(b) Modulated parasystole

The theoretical model for modulation of an ectopic ventricular pacemaker by the sinus rhythm developed by Moe *et al.* (1977) is quite close to the theoretical model for the periodically stimulated heart-cell aggregates outlined in §2. The normal sinus pacemaker is analogous to the microelectrode, and the ectopic focus is analogous to the spontaneously beating aggregate of heart cells. However, sinus beats which fall after an ectopic beat are blocked and consequently the sinus beat following an ectopic beat does not act to phase reset the ectopic rhythm. In addition, ectopic beats which fall during the refractory time of the ventricles are not observed (they are concealed).

We assume that the sinus rhythm acts to reset the ectopic focus, and call ϕ_i the phase of the ith sinus beat in the ectopic cycle. Assume that the ith sinus beat acts to phase reset the ectopic cycle. Then we expect that the phase of the next sinus beat will be at the phase $g(\phi_i) + \tau$ where $\tau = t_s/t_e$. However, if $g(\phi_i) + \tau > 1$ and if also $1 - g(\phi_i) > \theta/t_e$ then there will be an ectopic beat before the next sinus beat and the next sinus beat will not lead to a phase resetting. From the above, it can be shown that the only sinus beats that do not lead to phase resetting occur in the interval $0 < \phi < \tau - \theta/t_e$. Thus, the finite difference equation for modulated parasystole can be written

$$\left. \begin{aligned} \phi_{i+1} &= \phi_i + \tau, \quad 0 < \phi_i \leqslant \tau - \theta/t_e, \\ \phi_{i+1} &= g(\phi_i) + \tau \, (\text{mod } 1), \quad \tau - \theta/t_e < \phi_i \leqslant 1. \end{aligned} \right\} \tag{2}$$

This is equivalent to the formulation by Ikeda *et al.* (1983). For the special situation in which there is no phase resetting of the ectopic cycle $g(\phi) = \phi$, and the model is identical to the model for pure parasystole. If each sinus beat were effective in phase resetting the ectopic rhythm, the model would be identical to the model for periodically stimulated heart cells, except not every action potential of the heart cells would be observed.

It is straightforward to iterate (2) to determine the expected dynamics for a given function g. Such computations have been carried out with a number of different functional forms for g. Because of the compensatory pause, the sinus beat following an ectopic beat does not lead to a phase resetting of the ectopic rhythm and consequently the finite-difference equations for modulated para-systole can display discontinuities (see fig. 3 of Ikeda *et al.* 1983). Further the observation or non-observation of ectopic beats depends sensitively on the re-fractory time. As a consequence of these technicalities, the mathematical analy-sis of modulated parasystole presents greater difficulties than the analysis of entrainment of the chick heart cell aggregates or pure parasystole. Despite the difficulties of a general theory some observations can be made. For phase resetting curves measured experimentally, the effects of a stimulus in the immediate after-math of an action potential are negligible. This is so, for example, for the phase resetting for the chick heart cell aggregates. For such circumstances, there is

expected to be close correspondence between the entrainment zones using either
(1) or (2) because $g(\phi) = \phi$, for small values of ϕ. Thus in such circumstances there
will be zones of entrainment of the ectopic oscillator similar to the zones of the
Arnol'd tongues observed in figure 3 (see fig. 5 of Moe *et al.* 1977). However,
whether or not an ectopic beat will be observed is parameter-sensitive. Thus in
the 2:1 zone it is possible to observe no ectopic beats, or alternations of sinus
and ectopic beats (bigeminy). Similarly, in the 3:1 zone, one can observe either
no ectopic beats, or periodic sequences in which two sinus beats are followed by an
ectopic beat (trigeminy). Furthermore, because of non-monotonicity of g at some
stimulation strengths, the mathematical model for modulated parasystole is also
capable of displaying chaos (see also Ikeda *et al.* 1983).

   In view of the above considerations, it is not practical to give a complete analysis
of the dynamics of modulated parasystole. However, to illustrate some of the
properties of modulated parasystole we show results from a simulation using a
phase resetting curve obtained from the chick heart cell experiments. The use of
such a curve for modelling purposes is justified in view of the similarities between
phase resetting behaviour in the chick heart cell aggregates and in clinical data
(Jalife *et al.* 1982; Nau *et al.* 1982; Castellanos *et al.* 1984). Such a curve may be
more appropriate than the piecewise linear functions used by other workers (Moe
et al. 1977; Swenne *et al.* 1981; Ikeda *et al.* 1983). The results of the calculations
are shown in figure 7. We show the allowed values for the number of sinus beats
between ectopic beats. For some parameter values, the allowed values for the

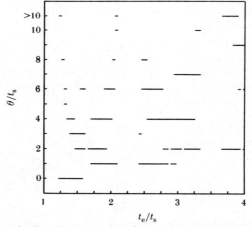

FIGURE 7. Allowed values for the number of sinus beats between ectopic beats for modulated
   parasystole from (2). Horizontal bars show the range of values of t_e/t_s for which a given
   value for the number of intervening sinus beats can be found. The narrow gaps in the
   horizontal bars correspond to stable phase locking zones in which a particular value for the
   number of intervening sinus beats is not found. A histogram in the same format as figure 5
   is not possible because the curves are jagged, where the degree of jaggedness depends on
   the fineness of the step size of the abscissa. The simulation used the same function as the
   phase resetting of chick heart cells with $A = 0.02$ and $\theta/t_s = 0.4$.

number of sinus beats between ectopic beats still obey the rules for 'pure' para-systole. However, there are also regions in which there are no ectopic beats observed. This is due to the phase locking of the ectopic oscillator to the sinus oscillator in such a fashion that all the ectopic beats fall in the refractory time following a sinus beat. Curves that give the probabilities for expected numbers of sinus beats between ectopic beats (as in figure 5) are extremely jagged for the parameter values in figure 7. This jaggedness depends on the fineness of the iteration.

The above analysis shows that the theoretical model for modulated parasystole, which has been developed by cardiologists and basic scientists with physiological and clinical data, can be cast as a problem about the bifurcations of circle maps. Because these maps are not necessarily invertible, and can be discontinuous, a challenging set of problems for mathematicians arises.

(c) *Variation of parameters*

Until now, we have only considered a few different ways that parameters can vary. The particular sorts of fluctuations that have been considered are motivated by the clinical records that will be discussed in §4, and also by known physiological mechanisms.

Although the sinus rhythm is frequently considered to be regular, all quantitative studies of the sinus rhythm have shown a surprising richness of behaviour with striking variability (Kitney & Rompelman 1980; Kobayashi & Musha 1982; Pomeranz *et al.* 1985; de Boer *et al.* 1985). There is a normal modulation of the sinus frequency with respiration, the so-called respiratory sinus arrhythmia. As well, some studies show fluctuations at a frequency of about 0.1 Hz which are

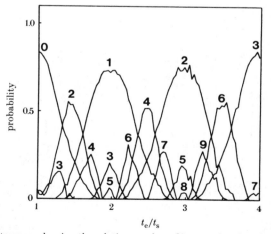

FIGURE 8. Histograms showing the relative number of intervening sinus beats between ectopic beats for pure parasystole with sinusoidal modulation of the sinus frequency. The modulation has a period of 5 t_s and an amplitude of 0.15 t_s with $\theta/t_s = 0.4$.

attributed to instabilities in the baroreceptor reflex (Kitney & Rompelman 1980). To assess the effects of sinus rate modulation we assume a sinusoidal modulation of the sinus rhythm.

We first consider the effects of sinusoidal modulation of the sinus rhythm during pure parasystole. Figure 8 shows the histograms showing the relative number of sinus beats between ectopic events as a function of t_e/t_s. In the region of the ratio, $t_e/t_s = 2$ and $t_e/t_s = 3$ new values not present for pure parasystole are found. In fact, the values for the number of sinus beats between ectopic events falls in the series $2n-1$ in the neighbourhood of the value $t_e/t_s = 2$ (i.e. the values are odd), and in the series $3n-1$ the neighbourhood of $t_e/t_s = 3$, where n is an integer. These rhythms are called concealed bigeminy and trigeminy respectively (Schamroth & Marriott 1963; Schamroth 1985).

Now consider the effects of sinusoidal modulation of the sinus rhythm during modulated parasystole in which the ectopic pacemaker is being reset. We consider an example in which the sinus rate modulation occurs in the 3:1 zone in which there is trigeminy (i.e. 2 sinus beats followed by an ectopic beat repeating periodically). Associated with the modulated sinus rate are shifts in the intervals from the sinus to the ectopic beats (R–X intervals) and the ectopic to the sinus beats (X–R intervals) which parallel the intervals between consecutive sinus beats (R–R intervals), figure 9. The shifts that are found parallel shifts observed in a clinical case of parasystole (see §4).

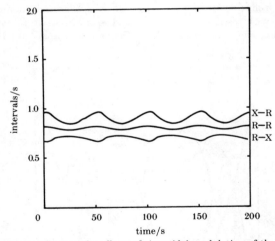

FIGURE 9. Time series showing the effects of sinusoidal modulation of the sinus rhythm in a mathematical model of modulated parasystole, equation (2), during trigeminy in which there are two sinus beats followed by an ectopic beat. The variations of the intervals between sinus beats (R–R), the interval from the sinus beat to the ectopic beat (R–X) and from the ectopic beat to the sinus beat (X–R) are shown. The modulation was assumed to have a period of 50 s and an amplitude of 0.16 s. The same phase resetting curve used for the chick heart cell simulations in figure 3 were used with $A = 0.046$. Other parameters are $t_s = 0.8$ s, $t_s/t_e = 0.265$, $\theta/t_s = 0.4$.

4. ANALYSIS OF HOLTER RECORDS

Previous studies of ECG records provide convincing demonstration that the mechanism of modulated parasystole is applicable in at least some circumstances (Jalife *et al.* 1982; Nau *et al.* 1982; Castellanos *et al.* 1984). However, arrhythmias in which there are frequent ventricular ectopic beats are extremely common in clinical practice and it is currently not clear the extent to which modulated parasystole will successfully account for the observed arrhythmias. Furthermore, detailed analysis of arrhythmias over extended periods of time is not generally attempted. We briefly discuss Holter recordings from two patients who display frequent ectopy.

First consider the ECG of an elderly patient who displayed long periods of intermittent ventricular trigeminy (figure 10*a*). This patient also had Cheyne–Stokes ventilation characterized by a regular waxing and waning of ventilation with a period of about 50 s. Holter records from this patient were obtained and digitized, and interbeat time intervals were measured. The time series reveals oscillations of all three component heart rate intervals (R–R, R–X, X–R) at the same frequency as the Cheyne–Stokes cycle (figure 10*b*). These shifts display the same phase relation as the theoretical model of modulated parasystole with slowly oscillating sinus frequency (figure 9). Periodic relations between heart rate,

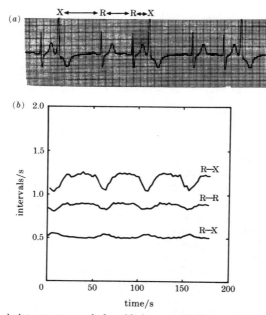

FIGURE 10. (*a*) Ambulatory ECG record of an elderly man with Cheyne–Stokes breathing showing episodes of ventricular trigeminy. (*b*) Time series showing the variation of the R–R, R–X and X–R intervals.

breathing and ventricular ectopy have been previously reported (Findley *et al.* 1984).

A second example is a middle aged patient with frequent ventricular ectopic beats (figure 11*a*). A 30 min record was printed on standard ECG paper at 25 cm s^{-1} and the intervals between the R-waves of sinus beats and ectopic beats were digitized. The number of consecutive sinus beats between ectopic beats fell in the

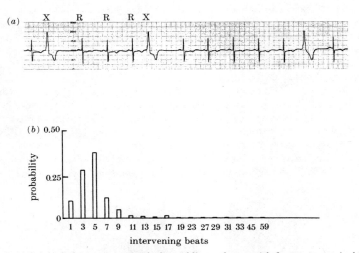

FIGURE 11. (*a*) Ambulatory ECG record of a middle-aged man with frequent ventricular ectopic beats. The number of sinus beats between ectopic beats over a 30 min period was always an odd number. This phenomenon is referred to as concealed bigeminy (Schamroth & Marriott 1963). (*b*) A histogram showing the relative numbers of intervening sinus beats between ectopic beats over a 30 min period.

range between 1 and 59. During this period the patient only displayed an odd number of sinus beats between ectopic events, i.e. there was concealed bigeminy (figure 11*b*). Figure 12 shows the consecutive values for the number of sinus beats between ectopic beats and also the sinus rate over a 30 min period.

A possible mechanism for this record is that there is a broad range of 2:1 entrainment between the sinus rhythm and the ectopic focus, but that some ectopic beats are blocked because of random fluctuation of the refractory time. If there are random fluctuations of the refractory time, then the probability for n sinus beats between ectopic beats decreases geometrically and is given by $p(1-p)^{\frac{1}{2}(n-1)}$, where n is an odd positive integer. However, an interesting feature of this record is that the histogram giving the number of sinus beats between ectopic beats is peaked around the value 5 (figure 11*b*), and this excludes a simple random flucturation of the refractory time.

An alternate hypothesis can be developed based on experimental studies in dogs. An electrical stimulus was delivered to the ventricles following every second

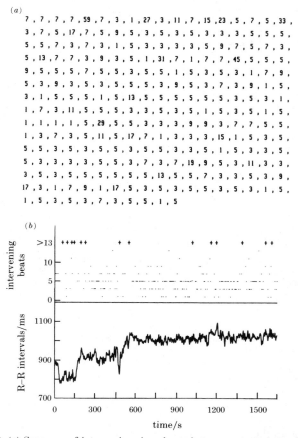

(a)
```
7 , 7 , 7 , 7 , 59 , 7 , 3 , 1 , 27 , 3 , 11 , 7 , 15 , 23 , 5 , 7 , 5 , 33 ,
3 , 7 , 5 , 17 , 7 , 5 , 9 , 5 , 3 , 5 , 3 , 5 , 3 , 3 , 3 , 5 , 5 , 5 ,
5 , 5 , 7 , 3 , 7 , 3 , 1 , 5 , 3 , 3 , 3 , 3 , 5 , 9 , 7 , 5 , 7 , 3 ,
5 , 13 , 7 , 7 , 3 , 9 , 3 , 5 , 1 , 31 , 7 , 1 , 7 , 7 , 45 , 5 , 5 , 5 ,
9 , 5 , 5 , 5 , 7 , 5 , 5 , 3 , 5 , 5 , 1 , 5 , 3 , 5 , 3 , 1 , 7 , 9 ,
5 , 3 , 9 , 3 , 5 , 3 , 5 , 5 , 5 , 3 , 9 , 5 , 3 , 7 , 3 , 9 , 1 , 5 ,
3 , 1 , 5 , 5 , 5 , 1 , 5 , 13 , 5 , 5 , 5 , 5 , 5 , 5 , 3 , 5 , 3 , 1 ,
1 , 7 , 3 , 11 , 5 , 5 , 5 , 3 , 3 , 5 , 3 , 5 , 1 , 5 , 3 , 5 , 1 , 5 ,
1 , 1 , 1 , 1 , 5 , 29 , 5 , 5 , 3 , 3 , 3 , 9 , 9 , 3 , 7 , 7 , 5 , 5 ,
1 , 3 , 7 , 3 , 5 , 11 , 5 , 17 , 7 , 1 , 3 , 3 , 3 , 15 , 1 , 5 , 3 , 5 ,
5 , 5 , 3 , 5 , 3 , 5 , 5 , 3 , 5 , 5 , 3 , 3 , 5 , 1 , 5 , 3 , 3 , 5 ,
5 , 3 , 3 , 3 , 3 , 5 , 5 , 3 , 7 , 3 , 7 , 19 , 9 , 5 , 3 , 11 , 3 , 3 ,
3 , 5 , 3 , 5 , 5 , 5 , 5 , 5 , 5 , 13 , 5 , 5 , 7 , 3 , 3 , 5 , 3 , 9 ,
17 , 3 , 1 , 7 , 9 , 1 , 17 , 5 , 3 , 5 , 3 , 5 , 5 , 3 , 5 , 3 , 1 , 5 ,
1 , 5 , 3 , 5 , 3 , 7 , 3 , 5 , 5 , 1 , 5
```

FIGURE 12. (a) Sequence of intervening sinus beats between ectopic beats for the patient in figure 11 over a 30 minute period. (b) The number of intervening sinus beats (upper) and the average R–R interval (lower) as a function of time. The R–R values represent a five-beat moving average.

sinus beat (Lee *et al.* 1974). It was found that even though the electrical stimulus was delivered at the same phase of the cycle (i.e. at a fixed delay) the effects were not the same; some stimuli were blocked whereas others were not. The interpretation of this finding was that the conduction of one ventricular stimulus increased temporarily the refractory time to subsequent stimuli. Simulations were carried out assuming the mechanism of modulated parasystole with a refractory period of the ventricles which is geometrically decreasing following an ectopic beat. With these assumptions it is possible to approximately reproduce the statistical features of the observed histograms if additional stochastic noise is added to the refractory time and a conduction delay is assumed. The simulations also showed that at faster sinus rates, an even number for the number of sinus beats between ectopic beats should be observed. In fact, in other portions of the record in which the R–R intervals were 700–800 ms, an even number of sinus beats between ectopic beats were occasionally observed. We hope to present a more complete analysis of this case in a subsequent publication.

From the above discussion it should be clear that a detailed analysis of dynamic data can be used to exclude plausible hypotheses about the underlying physiological mechanisms of these arrhythmias. However, it is extremely difficult to establish unambiguously the mechanism for the arrhythmias. Alternative hypotheses for these rhythms may also be consistent with the observed dynamics.

5. DISCUSSION

Simple biological and mathematical models of the intact heart display some features that can be found in clinically observed cardiac arrhythmias. This observation has implications both for basic science as well as clinical cardiology.

The simple model systems considered here are extreme caricatures of the anatomically and electrophysiologically complex human heart. A more complete mathematical model of the human heart must necessarily be formulated as nonlinear partial differential equations. We expect that the bifurcations and dynamics in these more realistic models should bear striking similarities to the bifurcations observed here.

Although we expect that the parasystolic mechanisms considered here are important in the generation of ventricular ectopy, other mechanisms such as reentry (Lee *et al.* 1974; Pick & Langendorf 1979) and delayed after depolarizations (Ferrier 1977; Wit *et al.* 1980) are also believed to be important. Consequently, it is likely that ventricular ectopic beats in any given individual may be due to one (or more) of several different mechanisms. Differential diagnosis of plausible mechanisms is difficult. Conventional analyses of ECGs that are now done, do not take into account the long-term fluctuations such as those presented in figure 12. An intriguing possibility is that nonlinear dynamics may eventually be useful in helping in diagnosing the mechanism and guiding the therapy of complex arrhythmias.

The human heartbeat shows striking fluctuations in rate during normal sinus rhythm and also during various arrhythmias. Although in some instances the fluctuations may be easy to characterize, more typically the dynamics are rich and highly complex. As an example, the sequences of the numbers of sinus beats between ectopic beats at first sight appear 'random' but contain regularities that reflect the underlying physiological mechanisms. The relative roles of 'stochastic noise' and 'deterministic chaos' in generating normal rhythms and arrhythmias are not clear. A full understanding will only be achieved from the integration of nonlinear mathematics with experimental physiology and clinical cardiology.

We have benefited from financial support from the Canadian Heart Foundation and the Natural Sciences and Engineering Research Council. We thank M. R. Guevara and J. Bélair for helpful conversations and J. Mietus for assistance. This work supported in part by a grant to A. L. G. from NASA Ames Research Center, Moffett Field, California.

REFERENCES

Arnol'd, V. I. 1965 *Am. math. Soc. Transl.* (2) **46**, 213–284.
Bélair, J. & Glass, L. 1985 *Physica D* **16**, 143–154.
Castellanos, A., Luceri, R. M., Moleiro, F., Kayden, D. S., Trohman, R. G., Zahman, L. & Myerburg, R. J. 1984 *Am. J. Cardiol.* **54**, 317–322.
Chung, E. K. 1977 *Principles of cardiac arrhythmias*. Baltimore: Williams and Wilkins.
Cvitanovic, P. (ed.) 1984 *Universality in chaos*. Bristol: Adam Hilger.
DeBoer, R. W., Karemaker, J. M. & Strackee, J. 1985 *Med. biol. Engng Comput.* **23**, 359–364.
DeHaan, R. L. & Fozzard, H. A. 1975 *J. gen. Physiol.* **65**, 207–222.
Ferrier, G. R. 1977 *Prog. cardiovasc. Dis.* **19**, 459–474.
Findley, L. J., Blackburn, M. R., Goldberger, A. L. & Mandell, A. J. 1984 *Am. rev. Resp. Dis.* **130**, 937–939.
Fleming, G. B. 1912 *Q. Jl Med.* **5**, 318–326.
Glass, L., Goldberger, A. L. & Bélair 1986a *Am. J. Physiol.* **251**, H841–H847.
Glass, L., Guevara, M. R., Bélair, J. & Shrier, A. 1984 *Phys. Rev. A* **29**, 1348–1357.
Glass, L., Guevara, M. R., Shrier, A. & Perez, R. 1983 *Physica D* **7**, 89–101.
Glass, L., Shrier, A. & Bélair, J. 1986b In *Chaos* (ed. A. V. Holden), pp. 237–256. Manchester University Press.
Goldberger, A. L., Bhargava, V., West, B. J. & Mandell, A. J. 1986 *Physica D* **19**, 282–289.
Goldberger, A. L., Findley, L. J., Blackburn, M. R. & Mandell, A. J. 1984 *Am. Heart J.* **107**, 612–615.
Goldberger, A. L., Findley, L. J., Blackburn, M. R. & Mandell, A. J. 1985 *Biophys. J.* **48**, 525–545.
Goldberger, A. L. & West, B. J. 1987 In *Chaos in biological systems* (ed. H. Degn, A. V. Holden & L. F. Olsen), New York: Plenum Press. (In the press.)
Guevara, M. R., Glass, L. & Shrier, A. 1981 *Science, Wash.* **214**, 1350–1353.
Guevara, M. R. & Glass, L. 1982 *J. Math. Biol.* **14**, 1–23.
Guevara, M. R., Shrier, A. & Glass, L. 1986 *Am J. Physiol.* **251**, H1298–H1305.
Ikeda, N., Yoshizawa, S. & Sato, T. 1983 *J. theor. Biol.* **103**, 439–465.
Jalife, J., Antzelevich, C. & Moe, G. K. 1982 *Pace* **5**, 911–926.
Jalife, J. & Moe, G. K. 1976 *Circulation Res.* **39**, 801–808.
Jalife, J. & Michaels, D. C. 1985 In *Cardiac electrophysiology and arrhythmias* (ed. D. P. Zipes & J. Jalife), pp. 109–119. Orlando: Grune & Stratton.
Katz, L. N. 1946 *Electrocardiology* (2nd edn). Philadelphia: Lea & Febiger.
Kaufman, R. & Rothberger, C. J. 1917 *Z. ges. exp. Med.* **5**, 349–370.
Keener, J. P. & Glass, L. 1984 *J. math. Biol.* **21**, 175–190.
Kinoshita, S. 1978 *Circulation* **58**, 715–722.
Kitney, R. I. & Rompelman, O. (eds.) 1980 *The study of heart-rate variability*. Oxford: Clarendon Press.
Kobayashi, M. & Musha, T. 1982 *IEEE Trans. Bio-med. Engng* **29**, 456–457.
Lee, M. H., Levy, M. N. & Zieske, H. 1974 *Am. J. Cardiol.* **34**, 697–703.
Lightfoot, P. 1978 *J. Electrocardiol.* **11**, 385–390.
MacKay, R. S. & Tresser, T. 1986 *Physica D* **19**, 206–237.
Mayer-Kress, G. (ed.) 1986 *Dimensions and entropies in chaotic systems*. Berlin: Springer-Verlag.
Mobitz, W. 1924 *Z. ges. exp. Med.* **41**, 180–237.
Moe, G. K., Jalife, J., Mueller, W. J. & Moe, B. 1977 *Circulation* **56**, 968–979.

Nau, G. J., Aldariz, A. E., Acunzo, R. S., Halpern, M. S., Davidenko, J. M., Elizari, M. V. & Rosenbaum, M. B. 1982 *Circulation* **66**, 462–469.
Pavlidis, T. 1973 *Biological oscillators: their mathematical analysis*. New York: Academic Press.
Perkel, D. H., Schulman, J. H., Bullock, T. H., Moore, G. P. & Segundo, J. P. 1964 *Science, Wash.* **145**, 61–63.
Phillips, J., Spano, J. & Burch, J. 1969 *Am. Heart J.* **78**, 171–179.
Pick, A. & Langendorf, R. 1979 *Interpretation of complex arrhythmias*. Philadelphia: Lea & Febiger.
Pomeranz, B., Macaulay, R. J. B., Caudill, M. A., Kutz, I., Adam, D., Gordon, D., Kilborn, K. M., Barger, A. C., Shannon, D. C., Cohen, R. J. & Benson, H. 1985 *Am. J. Physiol.* **248**, H151–H153.
Schamroth, L. 1980 *The disorders of the cardiac rhythm* (2nd ed.). Oxford: Blackwell.
Schamroth, L. 1985 In *Cardiac electrophysiology and arrhythmias* (ed. D. P. Zipes & J. Jalife), pp. 473–482. Orlando: Grune & Stratton.
Schamroth, L. & Marriott, H. J. L. 1963 *Circulation* **27**, 1043–1049.
Slater, N. B. 1967 *Proc. Camb. phil. Soc.* **63**, 1115–1123.
Smith, J. M. & Cohen, R. J. 1984 *Proc. natn Acad. Sci. U.S.A.* **81**, 233–237.
Swenne, C. A., de Lang, P. A., ten Hoopen, M. & van Hemel, N. M. 1981 *Comput. Cardiol.* 295–298.
van der Pol, B. & van der Mark, J. 1928 *Phil. Mag.* **6**, 763–775.
Wit, A. L., Cranefield, P. F. & Gadsby, D. C. 1980 In *The slow inward current and cardiac arrhythmias* (ed. D. P. Zipes, J. C. Bailey & V. Elharrar), pp. 437–454. The Hague: Martinus Nijhoff.

Chaos and the dynamics of biological populations

By R. M. May, F.R.S.

Department of Biology, Princeton University, Princeton,
New Jersey 08544, U.S.A.

As first emphasized in the early 1970s, the nonlinearities that are inherent in simple models for the regulation of plant and animal populations can lead to chaotic dynamics. This review deals with a variety of instances where chaotic phenomena can arise, particularly in interactions between prey and predators (including hosts and pathogens, hosts and parasitic insects, and harvested populations). Some of the complications in disentangling deterministic chaos from environmental noise will be discussed. The combination of population biology with population genetics leads to an even richer assortment of nonlinear phenomena and to the suggestion that many genetic polymorphisms may vary cyclically or chaotically (rather than being steady, as usually is assumed implicitly).

I argue that complex dynamics – including chaos – is likely to be pervasive in population biology and population genetics, even in seemingly simple situations. But superimposed environmental noise, in heterogeneous natural settings, will usually complicate the dynamics, making it unlikely that population data will exhibit elegant properties (such as universalities in period doubling) associated with the underlying maps. The existence of chaotic régimes of dynamical behaviour can, however, invalidate standard techniques for analysing population data to reveal density-dependent mechanisms; this, I believe, may currently be the most significant implication of dynamical chaos for population biology.

1. Introduction

A central task for population biologists is to disentangle, from the superimposed fluctuations caused by environmental noise and other chance events, the underlying mechanisms that regulate natural populations so that no one species of plant or animal increases without bound. Such studies lead us to consider simple equations that might describe the dynamics of natural populations if environmental noise and heterogeneity could be stripped away. A clear understanding of the dynamics of these simple and deterministic, but nonlinear, models then serves as a point of departure for evaluating the effects of various kinds of complications associated with environmental unpredictability and heterogeneity.

As is by now well known, such investigations in the early 1970s led to the realization that the simplest nonlinear models for populations with discrete, non-overlapping generations (first-order difference equations with one critical point) could exhibit a surprising array of dynamical behaviour (May 1974, 1976; Li & Yorke 1975; May & Oster 1976). Subsequent work showed that even richer dynamical behaviour could be generated by simple, deterministic equations for single populations with discrete but overlapping generations (higher-order difference equations), for single populations with continuous growth where regulatory effects contain time lags (time-delayed differential equations), and for two or more interacting populations. The dynamical properties of these models have been the subject of several recent reviews (Rogers 1981; Olsen & Degn 1985; Kloeden & Mees 1985; Lauwerier 1986 a, b; May 1983, 1986). Section 2 therefore does not attempt a comprehensive review, but rather is a guide to the existing literature with selective emphasis on a few points that are new or are not widely appreciated.

Given that chaos arises in the simplest equations propounded by 'muddy-boots' ecologists as natural descriptions of the underlying dynamics of their insect, fish, pathogen or other populations, the question of to what extent are chaotic dynamics actually observed arises. Section 3 summarizes recent studies of this question. My conclusion is that, in controlled laboratory settings, the array of dynamics from stable points, to stable cycles, to chaos can be seen, but that even in these artificial situations one cannot hope to see fine details of period doubling and the like (as one arguably can in some physical contexts, such as the onset of turbulence). In the natural world, the role of nonlinear phenomena (including possibly chaos) in the dynamics of many infectious diseases of humans and other animals is being understood in an increasingly explicit way. But for most natural populations, I believe environmental noise and other complications make it difficult to find examples of time series that show period doubling, intermittency, transitions to chaos, and other dynamical features that are clearly exhibited in some physiological and biochemical systems.

The fact that chaotic dynamics can arise from simple, density-dependent mechanisms does, however, have profoundly important implications for the way population biologists analyse data. Most existing work is based, usually implicitly, on the assumption that if density-dependent 'signals' could be dissociated from the confounding environmental noise, the population would be regulated to a steady, constant value. But if deterministic nonlinearities actually give chaotic time series which are effectively indistinguishable from stochastic fluctuations, the task of uncovering the regulatory signal can be much more complex. Some current work on this subject is reviewed in §4.

Ultimately, environmental noise does not act on populations as such, but on their constituent individuals. Thus we really need to derive deterministic models for the dynamics of populations from assumptions about the behaviour of individuals, so that the parameters in the population model derive from the biology of individuals. The effects of environmental noise can then be introduced in the proper way, through their effects on individuals. When this is done for insect populations in patchy environments, preliminary studies show that the interplay among nonlinear dynamics (giving rise possibly to cycles and chaos), spatial heterogeneity and environmental noise can invalidate standard techniques for detecting density-dependent mechanisms in natural populations. This work is also reviewed in §4, and it may represent the most significant implication that nonlinear dynamics holds for population biologists.

Section 5 extends the discussion to the dynamics of gene frequencies in populations where fitness functions are derived from ecological considerations (and thus can be frequency- or density-dependent). Again §5 is a brief outline of existing work. The conclusion is that the combination of population biology with population genetics can lead to a very rich assortment of nonlinear phenomena, with the implication that many genetic polymorphisms may vary cyclically or chaotically. Section 6 re-emphasizes the main messages in this review.

2. Chaotic dynamics in simple ecological models

2.1. *One-dimensional maps (single populations)*

Most readers will by now be familiar with the dynamical behaviour exhibited by the quadratic map,

$$x_{t+1} = ax_t(1-x_t). \tag{2.1}$$

If $3 > a > 1$, the fixed point at $x^* = 1 - 1/a$ is an attractor, and the system settles to the stable point made familiar by countless discussions in elementary mathematics courses. At $a = 3$ the system bifurcates, to give a cycle of period 2, which is stable for $1 + \sqrt{6} > a > 3$. As a increases beyond this, successive bifurcations give rise to a cascade of period doublings, producing cycles of periods 2, 4, 8, 16, ..., 2^n for a in the range $3.570.. > a > 3$. Beyond the point of accumulation of this cascade, $4 > a > 3.570..$, there lies an apparently chaotic régime, in which trajectories look like the sample functions of random processes. In detail, the apparently chaotic régime comprises infinitely many tiny windows of a-values, in which basic cycles of period k are born stable (accompanied by unstable twins), cascade down through their period-doublings to give stable harmonics of periods $k \times 2^n$, and become unstable; this sequence of events recapitulates the process seen more clearly for the basic fixed point of period 1. The details of these processes, and catalogues of the various basic k-cycles, have been given independently several times and are reviewed by May (1976), Collet & Eckmann (1980), and others.

The nature of the chaotic régime for such 'maps of the interval' is often misunderstood. In detail, the chaotic régime is largely a mosaic of stable cycles, one giving way to another with kaleidoscopic rapidity as a increases. But for essentially all practical applications, the chaotic region has the effectively random character that superficial inspection or numerical simulations suggest. This point is exemplified by the 'Lyapunov exponent' that is often computed as an index of chaotic behaviour. These exponents are analogous to the eigenvalues that characterize the stability properties of simpler systems. They are typically calculated by iterating difference equations, such as (2.1), and calculating the geometric average value of the slope of the map at each iterate: that is, for the difference equation

$$x_{t+1} = F(x_t), \tag{2.2}$$

the Lyapunov exponent λ is given by

$$\ln \lambda = \lim_{n \to \infty} \left\{ \frac{1}{n} \sum_{t=0}^{n} \ln \left(dF(x_t)/dx \right) \right\}. \tag{2.3}$$

For generically quadratic maps, there are unique attractors for most values of a in the chaotic régime. Therefore this calculation, if carried out exactly, or if the iterations are carried on long enough, will give values of λ less then unity ($\ln \lambda$ negative); an exact plot of the Lyapunov exponent for increasing a in the chaotic régime would be a hopeless jumble of ink lines, connecting the negative values (which arise for most values of a) to the set of positive values (which do have positive measure). But if only several tens of thousands of iterates are taken in numerical studies, $\ln \lambda$ is typically found to be positive in the chaotic régime (because transients take enormous times to die away for the very high-order cycles that predominate). Although inexact in a strictly mathematical sense, the 'chaotic' impression ($\ln \lambda > 0$) given by these numerical studies is probably more accurate for practical application than exact calculations ($\ln \lambda < 0$) would be! Thus the work on Lyapunov exponents reviewed by Olsen & Degn (1985) may be mathematically inaccurate (as pointed out by Gambaudo & Tresser (1983) and Kloeden & Mees (1985)), but it is usually correct in spirit.

The properties of exhibiting a stable point, or a cascade of period-doublings, or apparently chaotic dynamics are not peculiar to the quadratic map of (2.1), but are general to essentially all maps with one hump. Table 1 catalogues several such

TABLE 1. SOME FIRST-ORDER DIFFERENCE EQUATIONS, $x_{t+1} = F(x_t)$, TAKEN FROM THE BIOLOGICAL LITERATURE, WHICH CAN EXHIBIT CHAOTIC DYNAMICS

$F(x)$	source
$x \exp[r(1-x)]$	Moran (1950), Ricker (1954), Macfadyen (1963), Cook (1965), Pacala & Silander (1985)
$x[1+r(1-x)]$	Maynard Smith (1986), May (1972), Li & Yorke (1975)
λx, if $x < 1$; λx^{1-b}, if $x > 1$	Haldane (1953), Varley *et al.* (1973) and references therein
$\lambda x/(1+ax^b)$	Maynard Smith (1974), Bellows (1981)
$\lambda x(1+ax)^{-b}$	Hassell (1974)
$x[1/(a+bx)-\sigma]$	Utida (1957)
$\lambda_+ x$ if $x < 1$; $\lambda_- x$, if $x > 1$	Williamson (1974), with $\lambda_+ > 1$, $\lambda_- < 1$
$\lambda x[1-I(x)]$	May (1985), with $I(x)$ given by $1-I = \exp(-Ix)$
$\lambda x \, e^{-x} \sum_{i=0}^{\infty} \dfrac{x^i}{i!(1+\alpha i)}$	Pacala & Silander (1985), Crawley & May (1987)
$\dfrac{\lambda x}{(1+ax)^b + cx}$	Watkinson (1980)

first-order difference equations that have been proposed, in various theoretical and empirical contexts, as descriptions of biological populations. The basic mechanisms producing this array of behaviour in one-dimensional maps can, moreover, be understood in a very simple way, by using geometrical, combinatorial, or other approaches (May 1976). In particular, the generic process whereby period-doubling occurs, with a stable orbit of period $k \times 2^{n+1}$ appearing as the basic k-cycle harmonic of period $k \times 2^n$ becomes unstable, can be understood by a simple geometrical argument. This argument also gives an analytical estimate of the

Feigenbaum ratio as $\delta = 2(\sqrt{2}+1) \approx 4.83$ (obtained by approximating the period-$k \times 2^{n+1}$ map by a cubic in the neighbourhood of the fixed points of period $k \times 2^n$ at their transition from stability to instability: May & Oster (1980)). This is in good agreement (up to terms of relative order δ^{-2}) with the exact numerical computation $\delta = 4.669\ldots$.

The 'above remarks apply to maps that are generically quadratic in the sense that they have negative Schwarzian derivative (Singer 1978). In the absence of this restriction, we could have one-hump maps with, for example, a narrow range of attraction around the fixed point, but with the rest of the map generating very long chaotic transients which perpetuated until, by chance, an iterate fell within the range of attraction and settled to the fixed point. Such a system essentially has two different states, one stable state and one labile state (often called monostable); the example might be called monostable chaos. Alternatively, as discussed in more detail by Olsen & Degn (1985), it can be that the slope of the map in the neighbourhood of the fixed point is slightly below -1, with most of the rest of the map generating chaotic trajectories. The fixed point is then only weakly unstable, so that it takes many iterations to leave its neighbourhood; once the iterate has, however, left this neighbourhood, it can abruptly become fully chaotic, only to get caught in the quasistable neighbourhood of the fixed point again, sooner or later. Such alternation between an almost stationary state and chaotic fluctuations, repeated at apparently random intervals, is called intermittency. Intermittency does not arise for any of the maps listed in table 1, nor have population biologists given much thought to the phenomena. Its possible relevance in ecological contexts deserves more attention.

2.2. *A speculation about the history of the subject*

Given that simple equations, which arise naturally in many contexts, generate such surprising dynamics, it is interesting to ask why it took so long for chaos to move to centre stage the way it has over the past ten years or so. I think the answer is partly that widespread appreciation of the significance of chaos had to wait until it was found by people looking at systems simple enough for generalities to be perceived, in contexts with practical applications in mind, and in a time when computers made numerical studies easy.

Individually, the first two of these conditions were met long ago. Thus Poincaré found strange attractors in his studies of planetary dynamics, and he appreciated their significance, but these applications were sufficiently complicated that each could appear *sui generis*. Many concluded that, in nonlinear systems, each application is special, with no general messages. Even Lorenz's (1963) beautiful example of chaos in a simple system of three ordinary differential equations is complicated enough to have resisted a fairly full analysis until relatively recently. First-order difference equations, such as those in table 1, are indeed simple enough for a fairly complete understanding of their range of behaviour to be obtained, and several people (starting with Myrberg (1962) and Sharkovsky (1964)) did just that. But these earlier investigators were primarily interested in the exquisite mathematics, and do not seem to have had any messianic sense of the wider implications of their work.

In the 1940s and 1950s, several population biologists studied simple difference equations as models for practical problems: Moran (1950) in a entomological context; Ricker (1954) as a model for recruitment in fisheries. Their numerical studies uncovered stable points, stable cycles and even chaos. These people, however, were mainly interested in stable solutions, and they did not pursue the chaotic dynamics they found. Perhaps they distrusted the chaotic trajectories, as possible artifacts of their mechanical calculators.

In the studies of Moran, Ricker and other population biologists one had the conjunction of simple systems being studied with practical problems in view, but not my third conjectured ingredient of fast and reliable computers. All three ingredients did come together in the early 1970s, in work motivated by simple problems in population biology.

2.3. A 'completely chaotic' example

One particular example, which might have stimulated work on chaotic dynamics at an earlier date had it been studied earlier, arises as a simple and natural model for an insect population with discrete, non-overlapping generations that is regulated by a lethal pathogen which spreads in epidemic fashion through each generation, before reproduction. This system has recently been studied by May (1985) (see also Rogers *et al.* 1986). If the population increases by a factor λ from generation to generation in the absence of the pathogen, the population in generation $t+1$, N_{t+1}, is related to that in generation t by

$$N_{t+1} = \lambda N_t[1 - I(N_t)]. \tag{2.4}$$

Here $I(N_t)$ is the fraction infected, and thus killed before reproducing. The Kermack–McKendrick (1927) equation may be used to find the total fraction infected when an epidemic spreads through a population of magnitude N_t:

$$1 - I = \exp(-IN_t/N_T). \tag{2.5}$$

Here N_T is the threshold population size, which depends on the transmissibility and virulence of the pathogen: if $N_t < N_T$, the epidemic cannot spread, and $I = 0$; if $N_t > N_T$, the epidemic can spread, so that $I \neq 0$ and the effective reproductive rate is below λ. The one-dimensional map generated by this model has no stable points, and no stable cycles. As illustrated in figure 1, the system is 'completely chaotic', with an invariant measure for all values of λ ($\lambda > 1$). For a more detailed discussion, see May (1985) and Rogers *et al.* (1986).

The models for regulation of host insects by parasitoids, which are discussed below in §2.6, were developed in the 1920s and 1930s. They are very similar in spirit to (2.4) for regulation by a pathogen, except they do possess stable points and/or simple cycles among their possible dynamical behaviour. I have previously speculated that, had (2.4) been studied earlier, its completely chaotic dynamics might have forced population biologists to acknowledge the existence of chaotic dynamics much sooner.

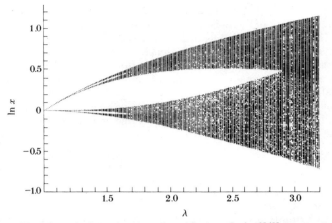

FIGURE 1. Plot of population values (on a logarithmic scale, $\ln(N/N_{\mathrm{T}})$) generated by iterating (2.4) many times, for each of a sequence of λ-values. The diagram gives an impression of the probability distribution of population values generated by this purely deterministic difference equation (after May 1985).

2.4. *Higher-dimension systems of difference equations*

The first-order difference equations listed in table 1 are metaphors for underlying density-dependencies in mechanisms regulating natural populations. Such regulatory effects, however, will often themselves contain time lags. This leads us to consider, as one of the simplest examples of such effects, the time-delayed logistic equation (Maynard Smith 1968):

$$x_{t+1} = \lambda x_t (1 - x_{t-1}). \tag{2.6}$$

Such an equation can be reformulated as a pair of first-order difference equations:

$$\left.\begin{array}{l} x_{t+1} = \lambda x_t (1 - y_t), \\ y_{t+1} = x_t. \end{array}\right\} \tag{2.7}$$

In this formulation, the dynamics may be followed by plotting pairs of points (x_t, y_t) in a two-dimensional phase plane; in the example (2.7), sensible trajectories are restricted to lie within the unit square (negative values x_t or y_t being taken to correspond to extinction). The system defined by (2.7) is a special case of more general classes of prey–predator relationships.

Equations (2.6) and (2.7) have been studied by Pounder & Rogers (1980), Aronson *et al.* (1982) and Rogers & Clarke (1981). Pounder & Rogers show the trajectories of points x_t, y_t are attracted to an invariant curve, which has an extremely complicated shape. Very briefly, this curve has infinitely many loops or folds issuing from the origin; the loops are successive images of the bottom arc of the curve. Once λ exceeds a critical value, the bottom arc of the curve can cross the x-axis, enabling the system to 'escape' to negative values (corresponding, biologically, to extinction). Interior loops of the invariant curve can also lead to 'escaping' trajectories, and extinction. The net result is that, for a range of λ-values, there are initial values which lead to extinction (possibly after very long times), while others can lead to cycles of high period or to chaotic fluctuations; the phase plane has a complicated filamentary structure, in which arbitrarily close initial points can undergo qualitatively different fates (exhibiting what Yorke has called 'fractal basin boundaries').

Rogers (1981) gives a review of this work, and Lauwerier (1986 b) has presented an exceedingly lucid analysis of a range of related two-dimensional systems, including a generalized form of (2.7):

$$\left.\begin{aligned} x_{t+1} &= \lambda x_t [1 - by_t - (1-b)x_t], \\ y_{t+1} &= x_t. \end{aligned}\right\} \tag{2.8}$$

Not surprisingly, these more general forms can have an even richer spectrum of dynamical behaviour than (2.7).

Such models lead into other two-dimensional systems, corresponding to interacting prey and predator populations with discrete, non-overlapping generations. One of the earliests such systems was propounded by the parasitologist Crofton (1971) as a description of certain kinds of host–parasite interactions:

$$\left.\begin{aligned} x_{t+1} &= \lambda x_t [1 + y_t]^{-k}, \\ y_{t+1} &= x_t y_t [1 + y_t]^{-k-1}. \end{aligned}\right\} \tag{2.9}$$

Preliminary analytical and numerical studies of this system by May (1979) reveal a fixed point, which attracts initial points lying along 'spiral arms' around it in phase space, with other initial points leading to high-order cycles or to apparent extinction. The above-described analyses of the delayed logistic equation help explain these earlier observations. Lauwerier's (1986 b) review extends to a variety of discrete prey–predator systems of this general kind, and he shows the complex dynamics and extreme sensitivity to initial conditions that can ensue.

2.5. *Host–parasitoid interactions*

Roughly 10 % of all metazoan species are insect parasitoids. These hymenopteran or dipteran species oviposit on or in their hosts (usually the egg, larval, pupal or adult stage of an insect, often a lepidopteran). Mathematical studies of the dynamics of such systems have gone hand-in-hand with empirical studies since the early work of Nicholson and Bailey in the 1930s (Hassell 1978). This is, in part, because each host produces either another host or a parasitoid in the next generation, which means that relatively realistic models have a simple structure

$$x_{t+1} = \lambda x_t f(y_t), \tag{2.10 a}$$

$$y_{t+1} = x_t - x_{t+1}/\lambda. \tag{2.10 b}$$

Here x and y represent the population densities of hosts and parasitoids, respectively. Each host either escapes parasitoid attack (with probability f, here assumed to depend only on y) and then produces λ progeny, or else is parasitized to produce one parasitoid in the next generation. A variety of forms have been propounded, in the entomological literature, for the 'search function' $f(y)$; some of these are reviewed by Lauwerier (1986 b), and to his list should be added the 'negative binomial' form $f(y) = [1 + ay/k]^{-k}$.

The dynamics of the system (2.10) will depend on λ and on one or more other parameters characterizing $f(y)$. In general, there can be a stable fixed point, which can undergo Hopf bifurcation to produce stable cycles. An interesting result which is little known among biological workers in this area, yet which explains the results of many numerical studies, is that the functional form of (2.10) generally implies the stable periodic solutions have approximate periods of at least six generations. This result is presented by Lauwerier (1986 b), and illustrated with numerical studies of several such systems (some of which have very complicated dynamics, including, for example, a stable inner seven-point cycle and an unstable outer cycle surrounding the unstable fixed point). Lauwerier's derivation, however, contains a minor error; Appendix A sketches the analysis.

2.6. *More general systems*

More generally, models for single populations can involve many discrete, but overlapping, age classes (see, for example, Levin & Goodyear 1980; Sparrow 1980; Hassell & May 1987). These eventually shade into situations where populations undergo continuous growth, but with time delays in recruitment or other processes, thus obeying time-delayed differential equations. The standard equations used by the International Whaling Commission to set quotas are of this kind, and can exhibit a rich range of period-doublings and chaotic behaviour as time lags lengthen and nonlinearities become steeper (although the parameters pertaining to real whale population produce only stable points). For reviews of this material, see May (1983) and Olsen & Degn (1985).

In short, simple and natural models for various kinds of biological populations exhibit cyclic and chaotic dynamics. We now ask whether population parameters will typically have values leading to such interesting dynamics.

3. CHAOS AND THE DYNAMICS OF REAL POPULATIONS

3.1. *Analysis of laboratory populations*

Single populations that are subject to density-dependent regulation, in a deterministic and homogeneous environment, are likely to be found only in the artificial setting of the laboratory. In this printed version of my lecture I shall not recapitulate my recent review of a variety of such laboratory studies (May 1986), several of which do indeed seem to show transitions from stable points to cycles, and possibly to chaos, as factors affecting demographic parameters are altered. Many of these laboratory investigations are, moreover, accompanied by explicit mathematical models for the dynamical behaviour.

Such laboratory studies are seen by some ecologists as unsatisfactory, in at least two respects. On the one hand, their artificiality may be argued to give them the status of living computers (conforming to unnaturally simple regulatory mechanisms, of little relevance to the dynamics of natural populations). On the other hand, for all the artificiality there remain many biological sources of noise, so that we do not in fact see period-doubling or other crisp transitions in dynamical régimes, but instead see at best fuzzy changes from constancy to cycles of increasing amplitude and/or irregularity, as biological parameters are varied.

3.2. *Natural populations: understanding the dynamics*

Interactions with other species mean that natural populations are usually governed by higher-order systems of equations. These complications are, of course, compounded by environmental noise and spatial heterogeneity. Thus, although broad patterns may be understandable (four-year cycles in many populations of small mammals in extremely seasonal environments may be an example), most work on the dynamics of natural populations is concerned just with trying to tease out density-dependent signals from a confusing background of density-independent noise, rather than with nonlinear details of the density-dependent signal as such.

The population dynamics of viral, bacterial, protozoan and helminth infections constitute one class of possible exceptions to this gloomy view. For one thing, the transmission of such organisms among hosts may be described more simply than is the case for the complex numerical and functional responses characterizing the population biology of most vertebrate prey–predator associations. For another, public health records afford long runs of data. Recent work on the nonlinear dynamics of host–pathogen associations seeks to explain the persistent and non-seasonal oscillations in the reported incidence of many childhood infections of humans in developed countries (measles, pertussis, rubella), and to predict temporal changes in incidence of infection following the implementation of specific vaccination programmes. These confrontations between nonlinear models for the dynamics, and population data, are also reviewed in my Croonian Lecture (May 1986). Period-doubling and chaos play a part in much of this work on non-seasonal periodicities (see, for example, Aron & Schwartz 1984; Grossman 1980).

3.3. *Natural populations: phenomenological analysis*

A very different approach to the analysis of population data has been pioneered by Schaffer & Kot (1985, 1986). The approach uses methods developed for physical problems by Packard *et al.* (1980), Takens (1981), and others, for situations where only one variable can be measured in a system possessing many independent variables. If some multidimensional attractor underlies the observed time series, it may be reconstructed (without any understanding of the fundamental mechanism that generates it) by choosing some fixed time lag, T, and plotting values of the variables $x(t)$, $x(t+T)$, $x(t+2T)$, ..., $x(t+[m-1]T)$ in m-dimensional space; the value chosen for T is not critical. The value of m is selected so that increasing its value by unity does not apparently result in any additional structure.

Schaffer & Kot (1985, 1986) have applied these techniques to the recorded numbers of cases of chickenpox, mumps and measles per month, $N(t)$, in New York and Baltimore before mass vaccination. They construct three-dimensional phase plots of $N(t)$, $N(t+T)$, $N(t+2T)$, with T fixed around two to three months. For all three-phase plots, Poincaré cross sections suggest the flows are indeed confined to a nearly two-dimensional conical surface, corresponding to some nearly one-dimensional map. Schaffer & Kot compute the Lyapunov exponents for these phenomenologically constructed one-dimensional maps, and find them all to be positive. Olsen & Degn (1985) review this work, and give a parallel but independent analysis of measles data from Copenhagen, which yields a one-dimensional humped map almost identical to those found for measles by Schaffer & Kot. Schaffer (1984) has also given a similar analysis of the Canadian data on apparent cycles in lynx abundance, arguing that this system also is chaotic and governed by a nearly one-dimensional map.

This phenomenological approach is clearly different in spirit from conventional approaches which seek to understand dynamical behaviour in terms of specific models based on underlying biological mechanisms. The approach holds the promise of providing new insights; its main problem is that it needs longer runs of data than are commonly available to population biologists.

4. NONLINEAR EFFECTS AND THE ANALYSIS OF POPULATION DATA

A growing amount of literature deals with the interplay between environmental noise and the intrinsic dynamics of nonlinear systems of the kinds discussed above. From a population biologist's point of view, one problem is that environmental fluctuations in reality affect individual organisms, and not population-level parameters as such. A fundamental analysis of how to extract density-dependent signals from environmental noise in population data therefore requires that we first understand how parameters characterizing the dynamics of a population derive from the behaviour of individuals.

Hassell (Hassell 1986; Hassell & May 1985) has recently explored these issues, both in illustrative but abstract models, and in relation to explicit data for populations of whitefly, *Aleurotrachelus jelinekii*, on viburnum bushes in England (Hassell *et al.* 1987). These theoretical and empirical studies exemplify the conjunction of three factors, which will influence the dynamics of most natural populations: in each generation, the overall population is distributed (often in a very non-uniform way) among many different patches; in each patch, density-dependent mechanisms affect the population dynamics (differently at different densities in different patches); and environmental fluctuations influence individual behaviour and thence population dynamics (again, possibly differently in different patches).

To begin, environmental fluctuations are ignored, and specific assumptions are made (or deduced from detailed observations) about the statistical distribution of the total population of reproductive adults among m distinct patches. Suppose next that each adult produces F offspring, and that the chance of each offspring surviving to the prereproductive dispersal stage, from a patch with i adults and thus iF offspring, is characterized by some density-dependent survival function $s(iF)$. The total population of reproductive adults in the next generation, N_{t+1}, is given by

$$N_{t+1} = m\{\sum_i p(i;N_t/m)\,s(iF)\,iF\}. \tag{4.1}$$

In this way we arrive at a first-order difference equation relating N_{t+1} to N_t, but now the parameters are those characterizing the distribution $p(\cdot)$ and survival $s(\cdot)$ at the patch and individual level. Note that regulatory effects are likely to be a combination of inter- and intragenerational effects, with some regulation occurring within each generation owing to density dependence acting differently at the different densities in various patches, and other regulatory effects deriving from between-generation differences in average population densities.

Such mathematical models can be used to generate pseudo-data (which can be noisy if the dynamics are chaotic), against which to test standard techniques of data analysis. Such studies show that conventional k-factor analysis, which in this case essentially plots changes in the total egg-to-adult mortality (the k-value on the y-axis) against initial adult density (plotted logarithmically on the x-axis), does reveal the density-dependent regulatory effects deriving from nonlinearities in the survival function $s(\cdot)$. This is true even though the analysis involves only average densities in successive generations, whereas much of the regulation occurs among patches within each generation.

The picture changes, however, when stochastic fluctuations are incorporated in the clutch size F, or in the parameters characterizing the dispersal and survivorship functions $p(\cdot)$ and $s(\cdot)$, respectively. Analysis of this theoretically generated pseudo-data by conventional k-factor analysis, applied to averaged densities in successive generations, in some cases still does reveal the underlying density-dependent effects, but in other cases does not. Whether or not the density-dependence that is actually present – though often predominantly acting within each generation – is revealed by such analysis depends on the magnitude of the stochastic fluctuations thus introduced (which is understandable), but also upon which parameter has been made noisy. May (1986) gives a much more explicit review of this recent work, complete with application to a model based on the whitefly-viburnum data.

The essential point is that when a population is distributed non-uniformly among many patches, with patches of different densities capable of exhibiting dynamical patterns ranging from stable points, through stable cycles, to chaos, the disentangling of density-dependent regulatory effects from superimposed environmental noise can be very difficult. The task may, indeed, often not be possible using conventional methods applied to overall average densities in successive generations. These questions go to the heart of the subject, calling for a reappraisal of conventional methods of gathering and analysing data.

5. POPULATION GENETICS AND CHAOTIC POLYMORPHISMS

However complex the dynamics in simple population models, things can get messier when population genetics is combined with population biology in models where fitness functions are frequency- or density-dependent.

The simplest such models deal with a single diallelic locus, with p_t and q_t being the relative proportions (or 'frequencies', $p+q=1$) of the two alleles A and a, respectively, in generation t. In a diploid population with random mating, the frequency of A in generation $t+1$, p_{t+1}, is related to that in generation t by a first-order difference equation

$$p_{t+1} = \frac{(p_t^2 W_{AA} + p_t q_t W_{Aa})}{(p_t^2 W_{AA} + 2p_t q_t W_{Aa} + q_t^2 W_{aa})}. \tag{5.1}$$

The quantities W_{ij} represent the fitnesses, or relative reproductive successes, of the three genotypes. If these fitnesses themselves depend on the gene frequency p_t, as can happen in a variety of biologically reasonable situations, we can have a highly nonlinear map, in the unit square, relating p_{t+1} to p_t.

In particular, May & Anderson (1983) have studied aspects of the coevolution of host-pathogen associations, using fitness functions derived from the kind of epidemiological considerations sketched in §2.3. Specifically, they assume in the simplest case that each of the three genotypes is susceptible to a particular pathogen (to which the other two are resistant), which spreads in epidemic fashion as described in §2.3. The fitness of genotype ij (ij $\equiv$ AA, Aa, aa) in generation t is then

$$W_{ij} = \lambda_{ij}[1 - \gamma_{ij}I(N_{ij})]. \tag{5.2}$$

Here λ_{AA} is the fitness, or relative productive success, of genotype AA in the absence of disease; γ_{AA} is the proportion of those infected who die; and $N_{AA} = N_t p_t^2$, with N_t the total density of the population in generation t. Similar definitions apply to the corresponding quantities for the genotypes Aa and aa. In each case, the Kermack–McKendrick (1927) relation gives the implicit expression (2.5) for $I(\cdot)$:

$$1 - I_{ij} = \exp[-I_{ij} N_{ij}/N_T]. \tag{5.3}$$

Here, as before, N_T is the threshold density for transmission of the infection.

Two cases can now be distinguished.

If overall population density in each generation is held constant by other ecological constraints, $N_t = K$, the proportion of each of the three genotypes to be infected – and thence, via (5.2), the fitness functions W_{ij} – depend only on the gene frequency, p_t. For such frequency-dependent (but not density-dependent) selection, gene frequencies in successive generations obey (5.1) with frequency-dependent W_{ij} from (5.2) and (5.3). The map has obvious fixed points at $(0,0)$ and $(1,1)$, but in general also has an interior fixed point by virtue of the propensity of p_t to increase from low values and decrease from high values (because disease spreads less effectively among rare genotypes and more effectively among common ones). The result can be a stable polymorphism, but for plausible values of the epidemiological parameters there can alternatively be cyclic or chaotic fluctuations in gene frequency.

If overall population density is itself regulated by the different diseases afflicting the three different genotypes, then both total population density and gene frequency can fluctuate cyclically or chaotically. Indeed, if all three diseases are lethal ($\gamma_{ij} = 1$ in all cases), it follows from §2.3 that the total population, and consequently the relative proportions of the genes A and a, have chaotic dynamics. Figure 2 illustrates such chaotic fluctuations in gene frequencies.

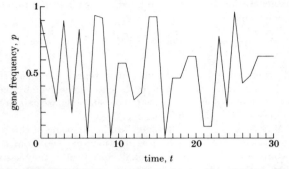

FIGURE 2. Illustrating the highly chaotic dynamical behaviour of the gene frequency p that can arise once the selective forces exerted by the pathogens are both frequency- and density-dependent. The figure plots successive iterates of p, as generated by the first-order difference equation (4.1) with (4.2) and (4.3) defining the fitness functions. After May & Anderson (1983), where details are given.

The above is a very brief summary of results presented in detail in May & Anderson (1983). The results have intrinsic mathematical interest, but their greater significance is that – like the results outlined in the previous section – they suggest we should think again about empirical aspects of biological studies. Given that many polymorphisms are thought to be maintained by frequency- or density-dependent mechanisms, it must be recognized that such maintenance is not necessarily at constant proportions of A and a; these nonlinear mechanisms can readily generate cyclic or chaotic fluctuations in gene frequencies. Whether such fluctuations are observed in natural populations is not known, because most studies have not reckoned with the possibility that gene frequencies may be continually changing, driven by their own chaotic dynamics.

6. DISCUSSION

In part, this review has aimed to present a range of models which provide deliberately oversimplified descriptions of the dynamics of natural populations of plants and animals. These simple, yet naturally derived, models can exhibit an astonishing array of dynamical behaviour.

I have argued that such behaviour does give qualitative insights into many population phenomena in the natural world, and that in some cases (dynamics of infectious diseases, for example) reasonably detailed understanding of nonlinear dynamical effects is emerging. But I doubt that phenomena like period-doubling or inverse Hopf bifurcations will be seen even in the cleanest population data, as they arguably are in physiological or biochemical contexts.

Recognition that density-dependent mechanisms can produce cyclic and chaotic behaviour in natural populations does, however, have important implications for the way certain kinds of data are analysed by ecologists. As clearly illustrated by applying conventional techniques of analysis to pseudo-data generated by models in which nonlinear dynamics, spatial heterogeneity, and environmental stoch-asticity roil together, the claim that a given set of data shows no evidence for density-dependence may often be a statement about the method of analysis, or about how the data were collected, and not about the biology of the system. Similarly, simple models suggest that cyclic or chaotic fluctuations in gene fre-quency can easily be maintained by selective forces whose magnitudes depend on gene frequencies or population densities. Again, there is need for a fresh look at gene frequencies in field populations, in the light of these results.

This work was supported in part by the National Science Foundation, under grant DMS-8604718.

Appendix A

This appendix establishes the result that host–parasitoid systems, as described by (2.10), are unlikely to have stable periodic orbits of order below 6.

Equation (2.10) has a fixed point at $y = y^*$, $x = x^* = y^*\lambda/(\lambda-1)$, where y^* is given by $\cdot \lambda f(y^*) = 1$. A linear stability analysis of this fixed point leads to a quadratic equation for the stability-determining eigenvalues σ:

$$\sigma^2 - \sigma(1+\alpha) + \lambda\alpha = 0. \tag{A 1}$$

Here I have, for convenience, defined

$$\alpha \equiv -x^*(df/dy)^*. \tag{A 2}$$

As f should be monotonically decreasing for increasing y, we should have $\alpha > 0$. The fixed point x^*, y^*, will be unstable as σ crosses the unit circle. At this Hopf bifurcation, we write $\sigma = e^{i\theta}$, and the imaginary part of (A 1) then gives

$$\sin(2\theta) - (1+\alpha)\sin\theta = 0. \tag{A 3}$$

That is, at the Hopf bifurcation from a stable point to a stable cycle, the phase angle is given by

$$\cos\theta = \tfrac{1}{2}(1+\alpha). \tag{A 4}$$

At the bifurcation, the parameters α and λ are related by $\lambda\alpha = 1$ (which can be obtained from the real part of (A 1) with $\sigma = e^{i\theta}$, or directly from the Schur–Cohn criterion). Using this to express α in terms of the more biologically familiar λ in (A 4), we finally arrive at

$$\cos\theta = \tfrac{1}{2}(1+1/\lambda). \tag{A 5}$$

As λ increases from around unity to very large values, the phase angle θ increases from 0° to 60°. We can thus obtain a six-point cycle in the limit $\lambda \to \infty$, but more generally a cycle of order roughly seven is the lowest likely to be found. This accords with several numerical studies presented by Olsen & Degn (1985), and with the earlier numerical work of Beddington *et al.* (1975). (Olsen & Degn's (1985) discussion has a sign wrong, to get $(1-1/\lambda)$ in (A 5); this leads them to conclude the phase angle θ lies between 60° and 90°, which would give cycles of approximate period between 4 and 6, rather than the correct 6 or more.)

REFERENCES

Aron, J. L. & Schwartz, I. B. 1984 Seasonality and period-doubling: bifurcations in an epidemic model. *J. theor. Biol.* **110**, 665–679.

Aronson, D. G., Chory, M. A., Hall, G. R. & McGehee, R. P. 1982 Bifurcations from an invariant circle for two-parameter families of maps of the plane: a computer assisted study. *Communs Math. Phys.* **83**, 303–354.

Beddington, J. R., Free, C. A. & Lawton, J. H. 1975 Dynamic complexity in predator–prey models framed in difference equations. *Nature, Lond.* **255**, 58–60.

Bellows, T. S. Jr 1981 The descriptive properties of some models for density dependence. *J. Anim. Ecol.* **50**, 139–156.

Collet, P. & Eckmann, J. P. 1980 *Iterated maps on the interval as dynamical systems.* Boston: Birkhauser.

Cook, L. M. 1965 Oscillation in the simplest logistic growth model. *Nature, Lond.* **207**, 316.

Crawley, M. J. & May, R. M. 1987 Population dynamics and plant community structure: competition between annuals and perennials. *J. theor. Biol.* **125**, 475–489.

Crofton, H. D. 1971 A quantitative approach to parasitism. *Parasitology* **63**, 179–193.

Gambaudo, J. M. & Tresser, C. 1983 Some difficulties generated by small sinks in the numerical study of dynamical systems: two examples. *Physics Lett.* A **94**, 412–414.

Grossman, Z. 1980 Oscillatory phenomena in a model of infectious diseases. *Theor. Popul. Biol.* **18**, 204–243.

Haldane, J. B. S. 1953 Animal populations and their regulation. *New Biol.* **15**, 9–24.

Hassell, M. P. 1974 Density dependence in single species populations. *J. Anim. Ecol.* **44**, 283–296.

Hassell, M. P. 1978 *The dynamics of arthropod predator–prey associations.* Princeton University Press.

Hassell, M. P. 1986 Detecting density dependence. *Trends Ecol. Evol.* **1**, 90–93.

Hassell, M. P. & May, R. M. 1985 From individual behaviour to population dynamics. In *Behavioural ecology* (ed. R. Sibly & R. Smith), pp. 3–32.

Hassell, M. P., Southwood, T. R. E. & Reader, P. M. 1987 The dynamics of the viburnum whitefly (*Aleurotrachelus jelinekii*): a case study on population regulation. *J. Anim. Ecol.* (In the press.)

Kermack, W. O. & McKendrick, A. G. 1927 A contribution to the mathematical theory of epidemics. *Proc. Roy. Soc., Lond.* A **115**, 700–721.

Kloeden, P. E. & Mees, A. I. 1985 Chaotic phenomena. *Bull. math. Biol.* **47**, 697–738.

Lauwerier, H. A. 1986a One-dimensional iterative maps. In *Chaos* (ed. A. V. Holden), pp. 39–57. Princeton University Press.

Lauwerier, H. A. 1986b Two-dimensional iterative maps. In *Chaos* (ed. A. V. Holden), pp. 58–95. Princeton University Press.

Levin, S. A. & Goodyear, C. D. 1980 Analysis of an age-structured fishery model. *J. Math. Biol.* **9**, 245–274.

Li, T.-Y. & Yorke, J. A. 1975 Period three implies chaos. *Am. math. Mon.* **82**, 985–992.

Lorenz, E. N. 1963 Deterministic nonperiodic flow. *J. atmos. Sci.* **20**, 130–141.

Macfadyen, A. 1963 *Animal ecology: aims and methods*, 2nd edn. London: Pitman.

May, R. M. 1972 On relationships among various types of population models. *Am. Nat.* **107**, 46–57.

May, R. M. 1974 Biological populations with nonoverlapping generations: stable points, stable cycles, and chaos. *Science, Wash.* **186**, 645–647.

May, R. M. 1976 Simple mathematical models with very complicated dynamics. *Nature, Lond.* **261**, 459–467.

May, R. M. 1979 Bifurcations and dynamic complexity in ecological systems. *Ann. N.Y. Acad. Sci.* **316**, 517–529.

May, R. M. 1983 Nonlinear problems in ecology and resource management. In *Chaotic behaviour of deterministic systems*, (ed. G. Iooss, R. H. G. Helleman & R. Stora), pp. 389–439. Amsterdam: North-Holland.

May, R. M. 1985 Regulation of populations with non-overlapping generations by microparasites: a purely chaotic system. *Am. Nat.* **125**, 573–584.

May, R. M. 1986 When two and two do not make four: nonlinear phenomena in ecology (the Croonian Lecture). *Proc. R. Soc. Lond.* B **228**, 241–266.

May, R. M. & Anderson, R. M. 1983 Epidemiology and genetics in the coevolution of parasites and hosts. *Proc. R. Soc. Lond.* B **219**, 281–313.

May, R. M. & Oster, G. F. 1976 Bifurcations and dynamic complexity in simple ecological models. *Am. Nat.* **110**, 573–599.

May, R. M. & Oster, G. F. 1980 Period doubling and the onset of turbulence: an analytical estimate of the Feigenbaum ratio. *Physics Lett.* **78**A, 1–3.

Maynard Smith, J. 1968 *Mathematical ideas in biology.* Cambridge University Press.

Maynard Smith, J. 1974 *Models in ecology.* Cambridge University Press

Moran, P. A. P. 1950 Some remarks on animal population dynamics. *Biometrics* **6**, 250–258.

Myrberg, P. J. 1962 Sur l'iteration des polynomes reels quadratiques. *J. Math. pures appl.* (9) **41**, 339–351.

Olsen, L. F. & Degn, H. 1985 Chaos in biological systems. *Q. Rev. Biophys.* **18**, 165–225.

Pacala, S. W. & Silander, J. A. 1985 Neighborhood models of plant population dynamics: I. single-species models of annuals. *Am. Nat.* **125**, 385–420.

Packard, N. H., Crutchfield, J. P., Farmer, J. D. & Shaw, R. S. 1980 Geometry from a time series. *Phys. Rev. Lett.* **45**, 712–716.

Pounder, J. R. & Rogers, T. D. 1980 The geometry of chaos: dynamics of a nonlinear second-order difference equation. *Bull. math. Biol.* **42**, 551–597.

Ricker, W. E. 1954 Stock and recruitment. *J. Fish. Res. Bd Can.* **11**, 559–623.

Rogers, T. D. 1981 Chaos in systems in population biology. *Prog. theor. Biol.* **6**, 91–146.

Rogers, T. D. & Clarke, B. L. 1981 A continuous planar map with many periodic points. *Appl. Math. Computat.* **8**, 17–33.

Rogers, T. D., Yang, Z. & Yip, L. 1986 Complete chaos in a simple epidemiological model. *J. math. Biol.* **23**, 263–268.

Schaffer, W. M. 1984 Stretching and folding in lynx fur returns: evidence for a strange attractor in nature? *Am. Nat.* **124**, 798–820.

Schaffer, W. M. & Kot, M. 1985 Nearly one dimensional dynamics in an epidemic. *J. theor. Biol.* **112**, 403–427.

Schaffer, W. M. & Kot, M. 1986 Differential systems in ecology and epidemiology. In *Chaos* (ed. A. V. Holden), pp. 158–178. Princeton University Press.

Sharkovsky, A. N. 1964 Coexistence of cycles of a continuous map of the line into itself. *Ukr. Math. J.* **16**, 61–71.

Singer, D. 1978 Stable orbits and bifurcation of maps of the interval. *SIAM Jl Appl. Math.* **35**, 260–267.

Sparrow, C. 1980 Bifurcation and chaotic behavior in simple feedback systems. *J. theor. Biol.* **83**, 93–105.

Takens, F. 1981 Detecting strange attractors in turbulence. *Lect. Notes Math.* **898**, 366–381.

Utida, S. 1957 Population fluctuation, an experimental and theoretical approach. In *Cold Spring Harb. Symp. Quant. Biol.* **22**, 139–151.

Varley, G. C., Gradwell, G. R. & Hassell, M. P. 1973 *Insect population ecology.* Oxford: Blackwell.

Watkinson, A. R. 1980 Density-dependence in single-species populations of plants. *J. theor. Biol.* **83**, 345–357.

Williamson, M. 1974 The analysis of discrete time cycles. In *Ecological stability* (ed. M. B. Usher & M. Williamson), pp. 17–33. London: Chapman and Hall.

Discussion

D. M. G. Wishart (*University of Birmingham, U.K.*). One of the most frequently analysed sets of data is the lynx–hare set. Is Professor May suggesting that we should, perhaps, declare a moratorium on this activity?

R. M. May. The lynx–hare data, compiled for the trading records of the Hudson Bay Trading Co., constitute one of the few long-term series available to population biologists. This is why it has been so much analysed and discussed (for a review of this body of work, see May 1980); we can ill afford a moratorium! Indeed, the most recent analysis, by Schaffer, uses Packard/Takens embedding methods, and is described in my paper.

Reference

May, R. M. 1980 *Nature, Lond.* **287**, 108–109.

J. BRAY (*House of Commons, U.K.*). Has Professor May thought theoretically about the epidemiology of AIDS?

R. M. MAY, F.R.S. Yes, and the preliminary work (which relates theory to epidemiological facts) is reported in Anderson & May (1986). This work, incidentally, provides a clear example where the interplay between heterogeneity (here, in levels of sexual activity within the population) and nonlinear dynamics makes it important first to do the dynamics and then to average. Analyses based on first averaging, attributing the same number of sexual partners to all individuals, do not fit observed patterns of rise in seroprevalence or other data.

Reference

Anderson, R. M. & May, R. M. 1986 *Phil. Trans. R. Soc. Lond.* B **314**, 533–576.

R. L. SMITH (*University of Surrey, U.K.*). I would like to draw attention to some statistical literature (e.g. Diggle and Gratton 1984) in which the idea of using simulation to fit a stochastic model to data has been developed. The simulations and data are used to construct a likelihood function, which may be thought of as a measure of closeness between the two, and parameters estimated by maximizing this function over the parameter space. Other measures, besides the likelihood function, are possible. The use of simulation described by Professor May seems based on a similar idea, and could perhaps be developed more fully in this way.

Reference

Diggle, P. J. & Gratton, R. J. 1984 Monte Carlo methods of inference for implicit statistical models with discussion. *Jl R. Statist. Soc.* B **46**(2), 193–227.

Fractal bifurcation sets, renormalization strange sets and their universal invariants

By D. A. Rand

*Mathematics Institute, Warwick University, Coventry CV4 7AL, U.K.†;
Applied Mathematics Program, University of Arizona, Tuscon,
Arizona 85721, U.S.A.*

The transition structure of the most common routes to chaos are organized by fractal bifurcation sets. Examples include the quasi-periodic transitions to chaos and the period-doubling structure found in Arnol'd tongues. In this paper I discuss the universality of such fractal bifurcation sets and their relation to strange invariant sets of renormalization transformations. An important result is that fractal bifurcation sets from within the same universality chaos are lipeomorphic. This implies that they have the same fractal structure and, in particular, the same Hausdorff dimension and scaling spectra. Some other invariants are introduced.

1. Introduction

The traditional renormalization formalism as invented to study phase transitions and as used in dynamical systems to study the universal properties of the transition to chaos relies upon finding a hyperbolic saddle point for a judiciously chosen transformation of some function space. Then the geometrical and dynamical structure of the saddle point and its stable manifold is used to deduce physically and mathematically interesting consequences. In this paper I shall discuss a more general situation which has a number of interesting applications to dynamical systems and which, I believe, is of even wider interest because it will have applications in other areas. In this generalization the role of the fixed point is played by a hyperbolic strange set Λ, which can be a strange saddle (e.g. a horseshoe) or a strange attractor. I call such sets *renormalization strange sets*. They are applied to deduce the structure and universality of complex fractal structures in parameter space.

In discussing various examples, one of my main aims will be to describe the sort of interesting universal objects that can be deduced from a renormalization strange set. For the traditional fixed point these are associated with the eigenvalues of the linearization at the fixed point and the functional structure of the fixed point. For the renormalization strange sets they will be dimensions and scaling functions and spectra derived from the detailed dynamical structure of the strange set Λ.

Dynamical systems often display complicated behaviour in parameter space. For example, if f_k is a one-parameter family of area-preserving twist maps of the annulus with f_0 integrable and f_1 having no homotopically non-trivial invariant

---

† Permanent address.

[45]

circles, then there is a very complicated cascade of breakdowns of these circles as k is increased from 0. The idea is to use renormalization strange sets to study and find universal properties of such complex processes. The scaling spectra mentioned above are invariants of the fractal structure and give a representation of the spectrum of scales in the fractal bifurcation sets and their relation with dynamical quantities. I shall discuss the following examples.

1. Critical circle maps with general rotation number which also give the universal structures associated with the breakdown of invariant tori in strongly dissipative systems.

2. The structure of the cascade of breakdowns of the invariant circles of area-preserving maps.

3. A theory for unimodal maps of the interval with general kneading invariant which generalizes the usual period-doubling theory.

4. An application to deduce self-similarity of the boundary of chaos in some 2-parameter unfoldings of homoclinic orbits.

The general idea of using a renormalization strange set to analyse the structures associated with general rotation number was first introduced in Ostlund *et al.* (1983). An early example in this area was the analysis of the structure of the boundaries of Seigel domains of general rotation number due to MacKay & Percival (1986). I will not discuss it here because it is similar to 1 and 2 and is more in the nature of a model problem and less physically interesting. Another early example that I will also not discuss is the application to the structure of quasiperodic Schrödinger operators due to Kohmoto *et al.* (1983) and Ostlund *et al.* (1982). This was perhaps the first example where a renormalization strange set was properly worked out and an excellent rigorous treatment is given by Casdagli (1986).

The ideas presented here are very speculative and much remains to be checked even in the various conjectures let alone the provision of rigorous proofs. Proofs in this area have tended to rely on computers to rigorously check certain estimates and although all the hypotheses I make can certainly be checked in this way in principle, at the present time it is beyond our computing facilities. However, I believe that the most important aspect of the sort of analysis presented here is that it provides a powerful mathematical picture which can be justified subject to a small number of precise hypotheses and which leads to a number of new and useful insights and conjectures about the transition to chaos which would not be seen without this picture.

2. Critical circle maps

A continuous map of the circle $\mathbb{T} = \mathbb{R}/\mathbb{Z}$ lifts to a map f of the universal cover $\mathbb{R}$ such that $f(x+1) = f(x)+1$. This lift is uniquely determined if one demands that $0 \leqslant f(0) < 1$. For x in $\mathbb{R}$ the rotation number of (f, x) is defined to be

$$\rho(f, x) = \lim \inf_{n \to \infty} n^{-1} (f^n(x) - x).$$

If f is a homeomorphism then the limit exists, is independent of x and depends continuously on f in the C^0-topology. This number $\rho(f)$ is called the *rotation number*. Now consider the following prototypical two-parameter family

$$f_{u, v}(x) = x + v - (\mu/2\pi) \sin 2\pi x.$$

If $|\mu| < 1$ then $f_{\mu,\nu}$ is a diffeomorphism and the dynamical structure is well understood thanks to the work of Arnol'd (1961), Herman (1979) and Yoccoz (1984). The reader is referred to Arnol'd's book (1984) for details. I am interested here in the critical case because little is known and it is of direct relevance to the structure of the breakdown of invariant tori in dissipative systems as shown in Rand (1984). They are in the same *universality class*. A circle map is called *critical* if it has a single critical point in $[0, 1)$ and if this is cubic. Thus each map in the family $f_\nu = f_{1,\nu}$ is critical. Now, if p/q is rational, let

$$I_{p/q} = \{\nu : f_\nu^q(x) = x + p\}.$$

Then each $I_{p/q}$ is a closed interval. Consider the complement

$$M_f = \{\nu : \rho(f_\nu) \text{ is irrational}\}.$$

Computer experiments by Jensen *et al.* (1983) indicate that this has zero Lebesgue measure and Hausdorff dimension $HD(M_f)$ approximately equal to 0.87. They also noted that $HD(M_f)$ appears to be independent of f. I will explain why this is so: if g_ν is a family close to f_ν then $HD(M_g) = HD(M_f)$. I will also introduce and explain universal functions associated with f which are universal and much better invariants.

The renormalization transformations for this case, invented by Ostlund *et al.* (1983) act on a space A of pairs of functions (ξ, η) of the form shown in figure 1 (also

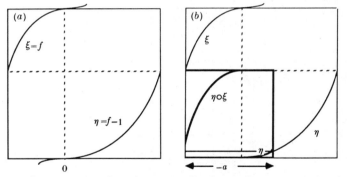

FIGURE 1. (*a*) Representation of a wide map as pair of commuting maps. (*b*) Construction of renormalized map of T_1.

see Feigenbaum *et al.* (1982) for a different, but related, transformation for the golden-mean case). In particular, they satisfy (*a*) $0 < \xi(0) = \eta(0) + 1$, (*b*) $\xi'(0) = 0$, (*c*) $\xi''(0) = 0$, (*d*) $\xi'(\eta(0)) \neq 0 \neq \eta'(\xi(0))$, (*e*) $\xi''(\eta(0)) \neq 0 \neq \eta''(\xi(0))$ and (*f*) $(\xi \circ \eta - \eta \circ \xi)^{(i)} = 0$ for $0 \leqslant i \leqslant 3$. Condition (*b*) gives criticality and implies $\eta'(0) = 0$ because of (*d*). The condition (*f*) is very important. In fact, the correct condition to impose is commutation, but the set of commuting pairs may not have a nice manifold structure so this weaker but adequate condition is used.

Note that if f is the lift if a critical circle map then the pair $u_f = (f, R_{-1} \circ f)$ satisfies these conditions. (R_α denotes the rotation $x \to x + a$.) Thus the critical circle maps may be regarded as embedded in A. Moreover, if $(\xi, \eta) \in A$ then the map $f = f_{\xi, \eta}$ defined by $f = \xi$ on $x \leqslant 0$ and $f = \eta$ on $x > 0$ defines a homeomorphism of the circle. Using this construction, define the rotation number $\rho(\xi, \eta)$ of the pair to be $\rho(f_{\xi, \eta})$. For each n in $\mathbb{N}$ define a renormalization transformation as follows:

$$T_n(\xi, \eta) = (a^{-1} \xi^{n-1} \circ \eta \circ a, a^{-1} \xi^{n-1} \circ \eta \circ \xi \circ a),$$

where $a = \xi^{n-1}(\eta(0)) - \xi^{n-1}(\eta(\xi(0)))$.

For T_n to be well defined each of the compositions must be compatible. This will be so if $1/(n+1) < \rho(\xi, \eta) < 1/n$. Therefore one takes the set D_n of pairs satisfying this condition for the domain of T_n. Then, because the D_n are disjoint, the T_n can be put together into a single transformation T whose domain is $\cup_{n \geqslant 1} D_n$ and which is defined by $T \mid D_n = T_n$.

Now T_n sends a pair with rotation number ρ to one with rotation number $\rho^{-1} - n$. Thus $T_n(D_n)$ contains pairs of every rotation number in $(0, 1)$. It is known from the rigorous work of Mestel (1985) that at the golden-mean fixed point one direction (roughly corresponding to ρ) is expanded and the rest are contracted. Thus, following Lanford (1987a, b), I assume the picture shown in figure 2. In particular, I assume that the geometry of the various intersections $T_n(D_n) \cap D_m$ is as shown and that there is uniform expansion in the one-dimensional 'vertical' direction and uniform contraction in the one-codimensional 'horizontal' direction.

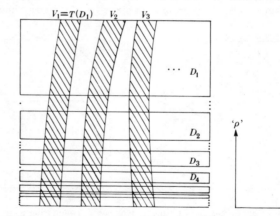

FIGURE 2. Schematic representation of the action of T on $\cup_n D_n$.

One deduces from this that A contains a T-invariant set Λ homeomorphic to the space $\mathbb{N}^{\mathbb{Z}}$ of biinfinite sequences of positive integers, and that the homeomorphism carries over the action of T on Λ to the shift σ on $\mathbb{N}^{\mathbb{Z}} : \sigma(\ldots a_{-1} a_0 a_1 \ldots) = \ldots b_{-1} b_0 b_1 \ldots$ where $b_i = a_{i+1}$. Given $\boldsymbol{a}^+ = a_1 a_2 \ldots$ in $\mathbb{N}^{\mathbb{N}}$, define

$$H_{a_1, \ldots, a_n} = \{u : T^{j-1} u \in D_{a_j} \quad \text{for} \quad 1 \leqslant j \leqslant n\}.$$

Then $H_{a_1, \ldots, a_{n+1}}$ is contained in $H_{a_1, \ldots, a_n}$ and

$$H_{\boldsymbol{a}^+} = \bigcap_{n \geqslant 1} H_{a_1, \ldots, a_n}$$

defines a 'horizontal' one-codimensional submanifold. Moreover, given $a^- = a_0 a_{-1} a_{-2} \ldots$ in $\mathbb{N}^{\mathbb{N}}$ define 'vertical' strips $V_{a_0 \ldots a_{-n}}$ inductively by

$$V_{a_0} = T(H_{a_0}), \quad V_{a_0, \ldots, a_{-n}} = T(V_{a_0, \ldots, a_{-(n-1)}} \cap H_{a_{-n}}).$$

Then $V_a = \cap_{n \geqslant 0} V_{a_0, \ldots, a_{-n}}$ is a 'vertical' one-dimensional curve. The homeomorphism between $\mathbb{N}^{\mathbb{Z}}$ and Λ sends $\ldots a_{-1} a_0 a_1 \ldots$ to $V_{a^-} \cap H_{a^+}$.

If $u \in H_{a^+}$ then $1/(a_j + 1) < T^{j-1}(u) < 1/a_j$ so $\rho = \rho(u)$ has continued fraction $[a_1, a_2, \ldots] = 1/(a_1 + 1/(a_2 + \ldots))$. Thus if u_ν is a one-parameter family close to Λ then

$$\rho(u_\nu) = [a_1, a_2, \ldots] \quad \text{if and only if} \quad u_\nu \in H_{a^+}$$

and the set $M_u = \{\nu : \rho(u_\nu) \text{ is irrational}\}$ corresponds to the intersection of the various H_{a^+} with the curve in Λ given by u_ν. The fact that the pairs in Λ and commuting pairs close to Λ come from analytic circle mappings follows from the glueing construction given in Rand (1984, 1987). From the theory of hyperbolic sets with one-dimensional unstable manifolds this implies that if u_ν is transverse to the H_{a^+}, then the fractal properties of the set M_u are the same as those for the families given by the V_{a^-} and the latter are independent of a^-. Thus these fractal properties are independent of u near Λ. It also follows that if f and g are two families transverse to the H_{a^+}s, then there is a lipeomorphism from M_f to M_g. A lipeomorphism is a Lipschitz homeomorphism that has a Lipschitz inverse. This can be tested numerically in the following way. For each $p/q \in \mathbb{Q}$ let $I_{p/q}(f) = [l_{p/q}(f), r_{p/q}(f)]$. For two critical families f and g plot the $l_{p/q}(f)$s and $r_{p/q}(f)$s against the $l_{p/q}(g)$s and $r_{p/q}(g)$s. This should give the graph $\gamma_{f,g}$ of a Lipschitz function. After I had conjectured this, Kim & Ostlund informed me that they have observed it in some numerical experiments (see Kim & Ostlund 1987). The graph they obtain for two families is shown in figure 3.

I now relate the above picture to the numerical work of Farmer & Satija (1984) and Umberger *et al.* (1986). Consider one-parameter families u_ν such that ν ranges from 0 to 1, $\nu \to \rho(u_\nu)$ is non-decreasing, $0 < \rho(u_\nu) < 1$, and $\rho(u_0) = 0$ and $\rho(u_1) = 1$. I call such families *full*. Each unstable manifold V_{a^+} can be parametrized by ν so

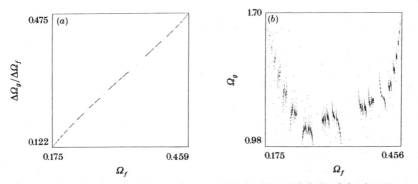

FIGURE 3. (a) The graph $\gamma_{f,g}$ obtained numerically by Kim and Ostlund for f as above and $g_\nu(x) = x + \nu - (5/16\pi)(\sin 2\pi x + 0.2 \sin 6\pi x)$. (b) The numerically computed derivative of the graph $\gamma_{f,g}$ of (a). (Courtesy of S. Kim.)

that it is such a family. Now given a full family u_ν define $u_\nu^{[n]}$ to be $T_n(u_{b+\nu(a-b)})$, where (a, b) is the interval of νs for which $u_\nu \in D_n$. Lanford (1985) has shown (i) that all the V_{a^+} can be parametrized so that

$$(V_{a_{-1}, a_{-2}, \dots})^{[a_0]} = V_{a_0, a_{-1}, a_{-2}, \dots}$$

in the obvious sense, and (ii) that if $u_\nu^{[a_0, \dots, a_{-n}]}$ is defined recursively as $(u_\nu^{[a_{-1}, \dots, a_{-n}]})^{[a_0]}$, then for an open set of full families u_ν,

$$u_\nu^{[a_0, \dots, a_{-n}]} \to V_{a_0, a_{-1}, \dots}$$

as $n \to \infty$, the convergence being uniform in $a_0, a_{-1}, \dots$ and exponentially fast in n. Farmer, Satija and Umberger had previously implemented the construction of the families $u_\nu^{[a_0, \dots, a_{-n}]}$ numerically and found reasonably good convergence and a representation of the V_a-s by using a projection into two representative dimensions.

In the above analysis I restricted to critical maps. This was not really necessary and I could have defined the transformation T on a neighbourhood of $\cup_n D_n$ in the space of all analytic pairs satisfying the conditions (a) and (c)–(f) above. This will introduce an extra unstable direction so that the stable manifolds have co-dimension two. In the extension to higher dimensional systems which include dissipative diffeomorphisms of the annulus $\mathbb{T} \times \mathbb{R}^{n-1}$ it is necessary to work in this way as the renormalization transformation $\tilde{T}$ there will act on pairs of mappings from $\mathbb{R}^n$ to $\mathbb{R}^n$ in a way similar to T (see Rand 1984) and for these maps there is no simple criterion corresponding to criticality. These maps contain Λ as they contain the singular maps whose image is one-dimensional as a subspace and $\tilde{T}$ is an extension of T. If the above conjectures about Λ are correct then from the methods in Rand (1984, 1987) it can be shown that Λ is also the appropriate renormalization strange set for $\tilde{T}$ because under renormalization the strongly dissipative maps and diffeomorphisms converge to singular one-dimensional maps. Suppose that $F_{\mu, \nu}$ is a two-parameter family of dissipative diffeomorphisms of the annulus

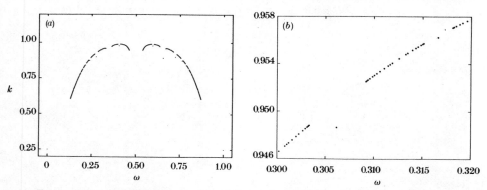

FIGURE 4. (a) A numerical approximation of the set of points (ω, k) at which the irrational circles break down for the mapping $(x, y) \to (x+y+\omega, \ \lambda y + (k/2\pi)\sin 2\pi(x+y))$ for $\lambda = 0.25$. (Courtesy of T. Bohr.) (b) A magnification of part of the set in (a).

such that the family of pairs $U_{\mu,\nu} = (F_{\mu,\nu},\ R_{-1} \circ F_{\mu,\nu})$ is transverse to the stable manifolds $(R_{-1}(x, y) = (x-1, y))$. By using methods similar to those in Rand (1984) for the golden-mean case it is possible to show that

$$K_F = \{(\mu, \nu) : U_{\mu,\nu} \text{ is in a stable manifold}\}$$

is precisely the set of points (μ, ν) at which the invariant circles break up. If $U_{\mu,\nu} \in H_{a_1, a_2, \dots}$ then the invariant circle of rotation number $[a_1, a_2, \dots]$ breaks up at (μ, ν). It follows from the hyperbolic structure of Λ that the set K_F will have the same fractal structure as the set M_f for transverse families of critical circle maps. In particular, $HD(K_F) = HD(M_f)$. Bohr has numerical evidence for this (see figure 4).

3. Breakup of the invariant circles of area-preserving maps

Area-preserving maps arise as return maps in hamiltonian systems with two degrees of freedom. Let f_k denote the prototypical family of area-preserving twist maps of the annulus given by

$$x_1 = x + y \pmod 1,$$
$$y_1 = y + (k/2\pi \sin 2\pi(x+y),$$

where $(x, y) \in \mathbb{T} \times \mathbb{R}$. By an invariant circle I shall mean a homotopically non-trivial circle C such that $f_k(C) = C$. If $k = 0$ every point lies on an invariant circle. If $0 < k < k_c = 0.9716354\dots$ then f_k has a Cantor set of invariant circles, whereas if $k > k_c$ there are none. At $k = k_c$ it is believed that there is a single circle whose rotation number is $gm = \frac{1}{2}(\sqrt 5 - 1)$. The breakdown of these circles is important because if they exist orbits are confined by them and therefore one has stability in the strongest sense that can be expected for such general non-integrable systems. Consequently the breakdown corresponds to the loss of confinement and stability.

To describe the cascade of breakdowns, I introduce the diagram K_f which consists of those (ω, k) such that if $0 < k' \leqslant k$ then $f_{k'}$ has an invariant circle of irrational rotation number ω. Below I will explain why it follows from the re-normalization picture that, up to lipeomorphism, the diagram K_f should be a universal object.

As for circle maps, instead of acting on single maps, the renormalization trans-formations act on pairs (ξ, η) of area-preserving maps. The basic idea is shown in figure 5. ξ acts on the curvilinear triangle to the left of l and η on that to the right. These area-preserving diffeomorphisms are embedded in the space B of such pairs by $f \to (f, R_{-1} \circ f)$ where $R_{-1}(x, y) = (x-1, y)$. Given a commuting pair (ξ, η) one obtains an annulus map by identifying $\xi(x)$ with $\eta(x)$ for each x in l and thus obtaining an annulus on which ξ and η define a diffeomorphism in a similar fashion to that of circle maps. Using this, one can define rotation numbers and invariant circles in the obvious way.

The basic renormalization transformations, which are essentially the same as those defined above for circle maps were first used by MacKay (1983, 1982). For

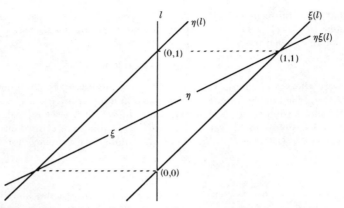

FIGURE 5. Schematic representation of the renormalization operation T, for area-preserving maps.

T_1 the basic idea is to delete everything to the right of $\eta(\xi(l))$ and then apply a coordinate change $B(x, y) = (ax, by - c - dx^2)$ to preserve the normalizations $\xi(0, 0) = (0, 0)$, $\eta(0, 1) = (0, 1)$ and $\xi(0, 1) = (1, 1)$ and make the picture look as before. That is

$$T_1(\xi, \eta) = (B^{-1} \circ \eta \circ B, B^{-1} \circ \eta \circ \xi \circ B).$$

This acts on rotation numbers as $\rho \to \rho^{-1} - 1$. More generally define

$$T_n(\xi, \eta) = (B^{-1} \circ \xi^{n-1} \circ \eta \circ B, B^{-1} \circ \xi^{n-1} \circ \eta \circ \xi \circ B)$$

with B defined as before. This acts on ρ as $\rho \to \rho^{-1} - n$. For our space B I take triples (ξ, η, ω) where (ξ, η) and ω are chosen so that $T_n(\xi, \eta)$ is well-defined if n is the integer part of $1/\omega$. Extend T_n to the triples as $(\xi, \eta, \omega) \to (T_n(\xi, \eta), \omega^{-1} - 1)$ which is defined on the subset D_n given by $1/(n+1) < \omega < 1/n$. In figure 6a I have represented the space of triples picking out the ω direction and one other expanding direction. Assume that these are uniformly expanding and that all the others are uniformly contracting, and also assume that the action of T_n on D_n is geometrically as that shown for T_2 on D_2. Independently, Escande (1987) and MacKay (1987) have worked out in detail a similar picture for an approximate renormalization scheme due to Escande and Doveil and I believe that they should be able to verify the above conjecture for this model.

The picture is a little complicated by the fact that one cannot isolate the renormalization strange set from the simple line of integrable maps given by $((x + y/\omega - \omega, y), (x + 1 + y, y))$ because circles with irrational rotation number break down arbitrarily close to the simple line. Given $\boldsymbol{a} = \ldots a_{-1} a_0 a_1 \ldots \in \mathbb{N}^{\mathbb{Z}}$, define $H_{\boldsymbol{a}^+} = \{u : T^{j-1} \in D_{a_j}\}$ as for circle maps. For this case of area-preserving maps it can be deduced from my picture that $H_{\boldsymbol{a}^+}$ is locally a one-codimensional submanifold with boundary (see figure 6). It is contained in $\omega = [a_1, a_2, \ldots]$. Let $\tilde{H}_{\boldsymbol{a}^+}$ be its boundary. Locally this will be a two-codimensional submanifold. The sets $H_{\boldsymbol{a}^+}$ and $\tilde{H}_{\boldsymbol{a}^+}$ are distinguished dynamically by the fact that if $u \in H_{\boldsymbol{a}^+} - \tilde{H}_{\boldsymbol{a}^+}$ then

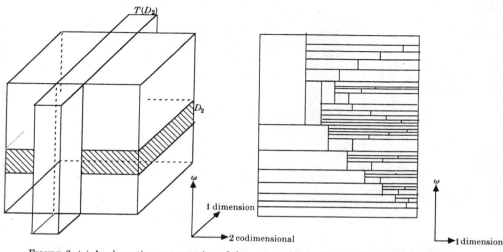

FIGURE 6. (*a*) A schematic representation of the action of T on D_2 in the space of triples for area-preserving maps. (*b*) A schematic representation of the projection of the invariant set obtained by factorizing out the stable directions.

$T^n u$ converges to the simple line as $n \to \infty$ whereas if $u \in \tilde{H}_{a^+}$ then it does not, instead it converges to the renormalization strange set Λ defined below. For rotation numbers satisfying a diophantine condition these two cases correspond to the existence of a smooth invariant circle and the existence of a critical invariant circle.

One can also define 'vertical strips' V_{a^-} as for circle maps. The intersection of the various $\tilde{H}_{a^+}$ and V_{a^-} gives the strange invariant set Λ. To get a better picture of Λ factor out the stable direction and only consider the two unstable directions. Then a picture as shown in figure 6*b* is obtained.

Now consider the two-dimensional surface Γ_f in B given by $\Phi(\omega, k) = (f_k, R_{-1} \circ f_k, \omega)$. If $\Phi(\omega, k) \in H_{a_1, a_2, \ldots}$ then $T^n(\Phi(\omega, k))$ converges to Λ or the simple line. Thus, by the arguments in Rand (1987), $\Phi(\omega, k)$ will have an invariant circle. Assuming that there is a fundamental neighbourhood U 'above' Λ such that each $(\xi, \eta) \in U$ has no invariant circles, then it follows that

$$K_f = \{(\omega, k) : \Phi(\omega, k) \in \tilde{H}_{a^+} \text{ for some } a^+\}.$$

Of course, if $\Phi(\omega, k) \in H_{a^+}$ then $\omega = [a_1, a_2, \ldots]$. Thus K_f can be identified with the intersection of Γ_f and the stable manifolds $\tilde{H}_{a^+}$ of Λ. Now suppose that g is close to f so that Γ_g is nearly parallel to Γ_f. Then by sliding along the $\tilde{H}_{a^+}$ one obtains a lipeomorphism from K_f to K_g (see figure 7). More generally, the following local result can be deduced from such an argument: for almost all $(\omega, k) \in K_f$ there is a neighbourhood U of (ω, k) and a lipeomorphism of $K_f \cap U$ onto a neighbourhood of the corresponding points in K_g. It should also follow from this that if Γ_f and Γ_g are transverse to the $\tilde{H}_{a^+}$s and $k_f(m)$ (resp. $k_g(m)$) is the infimum of those k such

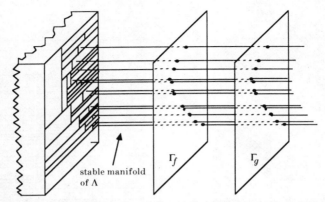

FIGURE 7. A schematic representation of the way in which one obtains a homeomorphism from K_f to K_g by sliding along the stable manifolds of the renormalization strange.

that the total Lebesgue measure of the invariant circles of f (resp. g) is m, then the graph consisting of the points $(k_f(m), k_g(m))$ is monotone and Lipschitz. Below in §6, I relate K_f to the so-called fractal diagram of Schmidt and Bialek and use this to define scaling spectra for this case.

Satija (1987) has carried out a numerical investigation of a set related to Λ. The viewpoint of her work is similar to that of Farmer & Satija (1984) on circle maps and its relation with the picture presented here is similar to the relation for circle maps described in §2.

4. UNIMODAL MAPS OF INFINITE DEPTH

A unimodal map f is an analytic map of the interval $I = [-1, 1]$ which has only one critical point c, that being a quadratic maximum. In fact, I will assume for simplicity that f is even and $c = 0$. Moreover, I shall normalize the maps and assume that $f(c) = 1$. The dynamics of f are largely determined by its kneading invariant ν_f which is defined in the following way. If $x \in I$ let $\theta_n(x)$ be $1, -1$ or 0 according as f^n is orientation preserving at x, reversing or $f^j(x) = x$ for some $0 \leqslant j < n$. Let $\boldsymbol{\theta}(x) = (\theta_0(x), \theta_1(x), \ldots) \in \{-1, 0, 1\}^{\mathbb{N}}$. Then the map $x \to \boldsymbol{\theta}(x)$ is monotone and the limit $\nu_f = \lim_{x \to 0^-} \boldsymbol{\theta}(x)$ exists in the product topology on $\{-1, 0, 1\}^{\mathbb{N}}$. The sequence ν_f is called the kneading invariant of f.

I am going to analyse the subset S of those f of *infinite depth*, i.e. S is maximal with the property that if $f \in S$ then there exists $n > 0$ and a subinterval J of I such that $f^n | J$ is conjugate to an element of S. The best known examples of maps f of infinite depth are those corresponding to the accumulation point of period-doubling. For these there is an interval J containing c such that $f^2 | J$ is conjugate to f. For the Feigenbaum fixed-point this conjugacy is actually linear.

Now I define some special finite kneading sequences. Let $\alpha(\frac{1}{2})$ denote the symbol $+$ and for each $n \geqslant 0$ inductively define $\alpha(2^n) = \alpha(2^{n-1})\overline{\alpha(2^{n-1})}$ where an overbar indicates that all signs are reversed. Thus $\alpha(1) = + -$ and $\alpha(2) = + - - +$. Now, if k is odd let $\alpha(2^n k) = \alpha(2^{n-1}) \overline{\alpha(2^{n-1}) \alpha(2^{n-1})} \alpha(2^n) \ldots \alpha(2^n)$ where $\alpha(2^n)$ is repeated $\frac{1}{2}(k-3)$ times. These sequences have the property that $\alpha(n) \alpha(n) \alpha(n) \ldots$ is the maximal admissible kneading invariant which is either periodic or antiperiodic with minimal period n.

Now for $k > 0$ odd, let D_k denote the set of those f such that (i) the first k terms of ν_f make up $\alpha(k)$ and (ii) $f^{2k}(c)$ lies between $-f^k(c)$ and $f^k(c)$, and (iii) f^{2j} does not lie between these if $j = 1, \ldots, k-1$. If $k = 2$ let D_2 denote the set of f such that (ii) and (iii) hold when $k = 2$.

These conditions imply that if J is the closed interval bounded by $-f^k(c)$ and $-f^k(c)$ then $f^k | J$ is a unimodal map on J. This I define a transformation T_k on D_k by

$$T_k(f) = a^{-1} f^k \circ a,$$

where $a = 1/f^k(c)$. Define an operator T on $\cup_{k \geqslant 2} D_k$ by $T | D_k = T_k$. Then I conjecture that one has horizontal and vertical strips and a similar picture to that for critical circle maps because the T_k act on kneading invariants in much the same way as the corresponding transformations there acted on rotation numbers. The stable manifolds H_{a^+}, $a^+ = a_1, a_2, \ldots, a_1, a_i \geqslant 2$, will consist of those f whose nonwandering set consists of an infinite number of hyperbolic repellors and a minimal attractor A whose dynamics are described in Jonker & Rand (1981). In particular, it follows from the renormalization that A can be described as follows: there exists a decreasing sequence of closed intervals J_m, $m \geqslant 0$, with $J_0 = I$ and such that if $l_m = a_0 \ldots a_{m-1}$ then $f^{l_m} | J_m$ is a unimodal map and if $J_{m,i} = f^i J_m$ for $i = 0, \ldots, l_m - 1$ then $A = \cap_{m \geqslant 1} \cup_i J_{m,i}$.

Because of the similarities with the analysis for circle maps I will not discuss this any further other than to mention that renormalization structures often allow one to lift results from one-dimensional singular systems to strongly dissipative diffeomorphisms. Because of this, I conjecture that for the Hénon map $H_{a,b}(x,y) = (a - x^2 - y, bx)$, if $|b|$ is sufficiently small then there is a Cantor set M in the a-axis such that if $a \in M$ then $H_{a,b}$ has a minimal attractor as for the corresponding unimodal map in the appropriate H_{a^+} and such that the fractal structure of M is independent of b. The idea that lies behind this is that one can define a similar renormalization transformation that acts on strongly dissipative diffeomorphisms and which will have the same renormalization strange set Λ because under renormalization the strongly dissipative diffeomorphisms converge to one-dimensional maps. When $|b|$ is sufficiently small the family $a \to H_{a,b}$ should be transverse to the stable manifolds of Λ.

5. SELF-SIMILARITY IN UNFOLDINGS OF HOMOCLINIC ORBITS

In this example I consider pairs of maps of the form $(\xi(x), \eta(x)) = (f(x \mid x \mid^\epsilon), g(x \mid x \mid^\epsilon))$, where f and g are analytic. Such maps arise naturally as the 'return maps' associated with homoclinic orbits (see Gambaudo *et al.* 1986). For example, consider a flow in $\mathbb{R}^3$ which (i) has the origin as a fixed point, (ii) the linearization at the origin is $(\dot{x}, \dot{y}, \dot{z}) = (\lambda_1 x, -\lambda_2 y, -\lambda_3 z)$ with each $\lambda_i > 0$ and (iii) if B denotes the box given by $|x| \leqslant 1, |y| \leqslant \frac{1}{2}, |z| \leqslant 1$ then the system is approximately linear inside B (scale the coordinates if not) and the two branches of the one-dimensional unstable manifold of the origin associated with the x-axis re-enters B in the face F given by $z = 1$. For example, this will be so if the system is a perturbation of one with a homoclinic orbit. One can define a return map from F to itself as the composition of the flow from F to the face $|y| = \frac{1}{2}$ given by the linear system and a map from the latter face to F along the unstable manifold. Provided that one restricts to a neighbourhood of the unstable manifold the latter map can be well-approximated by an affine map and, up to higher-order terms, the return map has the form $(x, y) \rightarrow (-\mu + bx^{\lambda_2/\lambda_1}, \tilde{\mu} + \tilde{b}yx^{\lambda_2/\lambda_1})$ on $x > 0$ with a similar expression for $x < 0$. The approximate map is therefore decoupled in the x-direction which is the interesting one for the dynamics because the other direction is contracting, and although this decoupling is only approximately the case as I have presented it, in many cases it can be shown to be rigorously true in some coordinate system because of the existence of a smooth foliation of strong stable manifolds. The return map can then be reduced to the action of a pair of maps (ξ, η) of the above form with ξ acting on $x < 0$ and η acting on $x > 0$.

Apart from the fact that ξ and η are not analytic (which is not important as they are derived from the analytic functions f and g), the main difference from the case of critical circle maps is that $\xi\eta(0) \neq \eta\xi(0)$. Thus the appropriate space of pairs is a neighbourhood of those with $\xi\eta(0) = \eta\xi(0)$. Because the pairs no longer define a homeomorphism of the circle there is not necessarily a single rotation number. Instead different points can have different rotation numbers for the same map. The set D_n is defined as the set of (ξ, η) such that every point has rotation number in $(1/(n+1), 1/n)$. Then the operator T is defined as for critical circle maps. The renormalization strange set Λ is constructed as in that case except that, because of the failure of $\xi\eta(0) = \eta\xi(0)$, there will be an extra unstable direction and consequently the stable manifolds will be two-codimensional. Moreover, it is not difficult to see that Λ and its stable manifolds will lie in the subspace given by $\xi\eta(0) = \eta\xi(0)$.

Now consider a two-parameter family of ordinary differential equations with fixed $\lambda_2 > \lambda_1$ which induce maps $u_{\mu,\nu} = (\xi_{\mu,\nu}, \eta_{\mu,\nu})$ such that $u_{\mu,\nu}$ is transverse to the stable manifolds. Then the set K_u of intersections corresponds to the set of pairs such that $\xi\eta(0) = \eta\xi(0)$ with irrational rotation number. Because there is an extra unstable direction there will be self-similar structure in the direction transverse to the curve given by $\xi\eta(0) = \eta\xi(0)$ which will scale in universal ways given by the scaling spectrum associated to Λ (see Gambaudo et al. 1986). Note that in this case a different universality class is obtained for each $\epsilon = \lambda_2/\lambda_1$. Also, for the application to differential equation one really needs to allow ϵ to vary, but this should not be difficult to handle.

Finally, I note that all this makes sense for small $\epsilon > 0$ a situation that has been analytically solved for the golden-mean circle case by Jonker & Rand (1983). It should be possible to treat the analogous case here.

6. Scaling and other spectra as universal invariants

As we have seen above one of the primary consequences of the existence of a renormalization strange set of the kind discussed is the universality of the associated fractal 'bifurcation' set up to lipeomorphism. The scaling spectrum is an invariant of this relation.

6.1. *The scaling spectrum*

For concreteness consider the renormalization strange set Λ constructed above for circle maps, although it will be clear that the results and approach is of much greater generality. Fix an unstable manifold $\Sigma = V_{a^-}$. Let $\Sigma_{a_1,\ldots,a_n}$ be the intersection of $H_{a_1,\ldots,a_n}$ with Σ. This is an open interval. Denote the set of such intervals by C_n. Let $N_n(a,b)$ denote the number of C in C_n such that, denoting Lebesgue measure by λ, $a < n^{-1} \lg \lambda(C) < b$, i.e. $\lambda(C) \in e^{n(a,b)}$. Let $S(a,b)$ be the growth rate $\lim_{n \to \infty} n^{-1} \lg N_n(a,b)$ and

$$S(\alpha) = \inf\{S(a,b) : a < \alpha < b\}.$$

Then $S(\alpha)$ is a continuous concave function whose Legendre transform is the function $P(\beta)$ which is the growth rate of the sums $\sum C_n \lambda(C)^{-\beta}$ (Bohr & Rand 1987; Gundlach 1986). $S(-\alpha)$ is Λs entropy function for characteristic exponents as defined in Bohr & Rand (1987). Moreover, because of the topological transitivity of Λ, S is independent of the particular V_{a^-}.

Now let u_ν be a full family as in §2 which is transverse to all the H_{a^+}s and so that $u_\nu^{[a_0,\ldots,a_{-n}]} \to V_{a_0,a_{-1},\ldots}$ as $n \to \infty$ as explained. Let $I_{a_1,\ldots,a_n}$ denote the set of ν such that u_ν lies in $H_{a_1,\ldots,a_n}$. Let $S_u(\alpha)$ be defined as $S(\alpha)$ except that the $I_{a_1,\ldots,a_n}$ replace the $\Sigma_{a_1,\ldots,a_n}$. Let $C_{n,\nu}$ denote the interval $I_{a_1,\ldots,a_n}$ containing ν. Then the following result follows from those of Lanford (1987a,b), Bohr & Rand (1987) and Gundlach (1986).

THEOREM. *If u_ν is transverse to the $H_{a^+}s$, then $S_u = S$. Hence* (i) *S_u is independent of u,* (ii) *S_u is real-analytic on its support,* (iii) *$D_u(\alpha) = -S_u(\alpha)/\alpha$ is the Hausdorff dimension of the set of points ν such that $\lim_{n \to \infty} n^{-1} \lg \lambda(C_{n,\nu}) = \alpha$,* (iv) *$HD(M_u)$ is the value of α such that $D'_u(\alpha) = 0$ and* (v) *the total size $\sum_k \lambda(C)$ of the complement of the gaps goes to zero like e^{nP} where*

$$P = S_u(-\alpha) - \alpha,$$

where α satisfies $S'_u(\alpha) = -1$.

The interpretation of this in terms of parameter space is as follows: Let u_ν be a critical family. Consider the set A_n of rational numbers of the form $[a_1,\ldots,a_n]$ where each $a_i \in \mathbb{N}$. The complement of the set of intervals $I_{p/q}, p/q \in \cup_{i-1}^n A_i$ consists of a set K of open intervals. These are precisely the $I_{a_1,\ldots,a_n}$. So the universal function tells us the set of scales in this set as $n \to \infty$. Alternatively, and perhaps more appealingly, the fact that each $I_{a_0,a_1,\ldots,a_n}$ contains the phase-locked tongue $I_{[a_0,\ldots,a_n]}$ which is mapped onto $I_{[a_1,\ldots,a_n]}$ by T_{a_0} can be used to deduce that the same result is true and the same spectrum obtained when the $I_{a_1,\ldots,a_n}$ are replaced by the $I_{[a_1,\ldots,a_n]}$.

6.2. *The q-spectrum*

The most interesting quantity associated with the intervals $I_{p/q}$ is the length q of the associated periodic orbits of the u_ν, $\nu \in I_{p/q}$. Thus it is interesting to construct an invariant which together with the geometry, takes account of this dynamical information. Let $N_n(a,b)$ be the number of $C = I_{[a_1,\ldots,a_n]}$ such that $a < \lg \lambda(C)/\lg q_n < b$, where $q_n = q_n(a_1, a_2, \ldots)$ is given by $[a_1,\ldots,a_n] = p/q_n$ in lowest-order terms. The functions $\tilde{S}(\alpha)$ is defined in terms of the $N_n(a,b)$ as before so that, roughly speaking, the number of $I_{[a_1,\ldots,a_n]}$ with $\lambda(I_{[a_0,\ldots,a_n]}) = q_n^\alpha$ grows like $e^{n\tilde{S}(\alpha)}$.

If the $I_{a_1,\ldots a_n}$s are used instead of the $I_{[a_0,\ldots,a_n]}$s the same function is obtained, and this in turn is independent of the choice of V_{a^-}. Thus, if u_ν is transverse to the H_{a^+}s, the function $\tilde{S}$ is independent of u.

This function is analysed by using the thermodynamic formalism of Collet & Lebowitz (unpublished results), Rand (1986) and Gundlach (1986). In the latter there is a general treatment for the fluctuation spectra S_{ψ_1,ψ_2} of quantities of the form $\sum_{i=0}^{n-1} \psi_2(T^i(x))/\sum_{i=0}^{n-1} \psi_1(T^i(x))$ where ψ_1 and ψ_2 are Hölder continuous. The function $\tilde{S}$ is the fluctuation spectrum associated with $\psi_1(x) = -\lg \| dT(x) | E^u(x) \|$, where $E^u(x)$ is the tangent space to $V_{a_0,a_{-1}\ldots}$ where $x = \ldots a_{-1} a_0 a_1 \ldots$ and $\psi_2(x) = [a_0 a_{-1} a_{-2}\ldots]$. Then ψ_1 controls the rate at which the lengths of the $I_{a_1,\ldots,a_n}$s decrease as $n \to \infty$ and $\sum_{i=0}^{n-1} \lg \psi_2(T^i(x))/\lg q_n(x)$ is bounded above and below by positive constants independent of x and n, so ψ_2 determines the increase in q_n as $n \to \infty$.

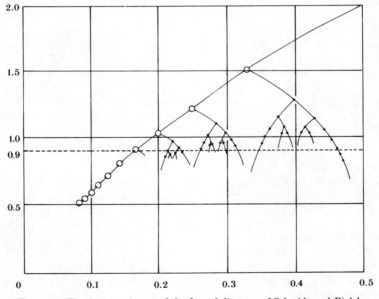

FIGURE 8. The first two layers of the fractal diagram of Schmidt and Bialek.
(Courtesy of G. Schmidt.)

It follows from these results that $\tilde{S}$ is a real-analytic concave function on its support. Therefore, if $\tilde{S}$ has a maximum it will be unique. Assume that this maximum is achieved at $\alpha = \alpha_{max}$. Then it follows from the definition of $\tilde{S}$ that for large n, almost all of the phase-locked intervals $I_{[a_0, \ldots, a_n]}$ scale as $q^{\alpha_{max}}$.

6.3. *Scaling spectra for area-preserving maps*

Scaling spectra can be defined for the set K_f defined in §3 for area-preserving maps and, as for the circle, these will be independent of f. In particular, the Hausdorff dimension of K_f should be universal. The set K_f should be approximated by the so-called fractal diagram of Schmidt & Bialek (1982). This is the set $\tilde{K}_f = \bigcup_{n \geqslant 1} \tilde{K}_{f,n}$ of points (ω, k) defined as follows: for each $\omega = [a_1, a_2, \ldots, a_n] \in A_n$ let $k_c(\omega)$ denote the value of k at which the minimax orbit with rotation number ω becomes destabilized i.e. bifurcates from elliptic to hyperbolic. Then $\tilde{K}_{f,n} = \{(\omega, k_c(\omega)): \omega \in A_n\}$. A picture of $\tilde{K}_{f,1} \cup \tilde{K}_{f,2}$ is shown in figure 8. This diagram will be self-similar because each T_n maps the subset of $\tilde{k}_{f,m}$ with $a_1 = n$ onto $\tilde{K}_{f,m-1}$. The set K_f should be the boundary of $\tilde{K}_f$, i.e. $K_f = \mathrm{cl}(\tilde{K}_f) - \tilde{K}_f$.

By using K_f a natural scaling spectrum is constructed as follows. For given $a_1, \ldots, a_n$ let $\omega_{n,m} = [a_1, \ldots, a_{n-1}, m]$ and $\delta_{n,m}$ be the vertical distance between $(\omega_{n,m}, k(\omega_{n,m}))$ and $(\omega_{n,m+1}, k(\omega_{n,m+1}))$ for $m \geqslant 0$. Let $N_n(a, b)$ denote the number of $a_1, \ldots, a_{n-1}, m$ such that $a < n^{-1} \lg \delta_{n,m} < b$ and let $S(a, b)$ be the growth rate of the $N_n(a, b)$ as $n \to \infty$. Then if $S(\alpha) = \inf\{S(a, b): a < b\}$ and $D_f(\alpha) = -S(\alpha)/\alpha$ then under the usual transversality hypothesis that Γ_f is transverse to the $H_{a'}$s or even locally so, $D_f(\alpha)$ is independent of the family f. It will have properties similar to those of the analogous functions defined in §6.1 above for critical circle maps.

I am indebted to the Applied Mathematics Program of the University of Arizona for its hospitality during the visit when this paper was written. This was partly supported by the University of Arizona and the US ONR under grant N00014-85-K-0412.

References

Arnol'd, V. I. 1984 *Geometrical methods in the theory of differential equations.* Berlin: Springer-Verlag.

Arnol'd, V. I. 1961 Small denominators. I. On the mapping of a circle into itself. *Izv. Akad. Nauk. SSSR Math. Ser.* **25** 21–86 (Am. math. Soc. Transl. (2) **46**, 213–284 (1965).)

Bohr, T. & Rand, D. A. 1987 The entropy function for characteristic exponents. *Physica* D **25**, 387–398.

Casdagli, M. 1986 Symbolic dynamics for the renormalisation map of a quasi-periodic Schrödinger equation. *Communs math. Phys.* **107**, 295–318.

Escande, D. F. 1987 *Renormalisation strange set for KAM tori.* (In preparation.)

Farmer, J. D. & Satija, I. I. 1984 Renormalisation of the quasi-periodic transition to chaos for arbitrary rotation numbers. *Phys. Rev.* A **31**, 3520–3522.

Feigenbaum, M. J., Kadanoff, L. P. & Shenker, S. 1982 Quasi-periodicity in dissipative systems: a renormalisation group analysis. *Physica* D **5**, 370–386.

Gambaudo, J.-M., Procaccia, I., Thomae, S. & Tresser, C. 1986 New universal scenarios for the onset of chaos in Lorenz-like flows. Preprint.

Greene, J. M., MacKay, R. S. & Stark, J. 1986 Boundary circles for area-preserving maps. Preprint. University of Warwick.

Gundlach, M. 1986 Large fluctuations of pointwise dimension, characteristic exponents and pointwise entropy in Axiom A attractors. M.Sc. Dissertation. University of Warwick.

Herman, M. 1979 Sur la conjugasion differentiable des diffeomorphismes du cercle a des rotations. *Publs math. I.H.E.S.* **49**, 5-234.

Jensen, M. H., Bak, P. & Bohr, T. 1983 Complete devil's staircase, fractal dimension and universality of the mode-locking structure in the circle map. *Phys. Rev. Lett.* **50**, 1637–1639.

Jonker, L. & Rand, D. A. 1981 Bifurcations in one dimension. 1. The nonwandering set. *Invent. Math.* **62**, 347–365.

Jonker, L. & Rand, D. A. 1983 Universal properties of maps of the circle with ε-singularities. *Communs math. Phys.* **90**, 273–292.

Kim, S. & Ostlund, S. 1987 Universal scalings in critical circle maps. (In preparation.)

Kohmoto, M., Kadanoff, L. P. & Tang, C. 1983 Localisation problem in one dimension: Mapping and escape. *Phys. Rev. Lett.* **50**, 1870–1872.

Lanford, O. E. 1987 Renormalisation group methods for circle mappings. In *Proc. 1985 Groningen Conference Statistical Mechanics and Field Theory: Mathematical Aspects.* (In the press.)

Lanford, O. E. 1987b Renormalisation analysis for critical circle mappings with general rotation number. In 8th *International Congress of Mathematical Physics, Marseilles, 1986* (ed. M. Mebkhout & R. Sénéor), pp. 532–536. World Scientific.

MacKay, R. S. 1982 Renormalisation in area-preserving maps. Ph.D. Thesis. University of Princeton.

MacKay, R. S. 1983 A renormalisation approach to invariant circles in area-preserving maps. *Physica* D **7**, 283–300.

MacKay, R. S. 1987 Exact results for an approximate renormalisation scheme. (In preparation.)

MacKay, R. S. & Percival, I. C. 1986 Universal small-scale structure near the boundary of Siegel disks of arbitrary rotation number. Preprint. University of Warwick.

Mestel, B. D. 1985 A computer assisted proof of universality for cubic critical maps of the circle with golden-mean rotation number. Ph.D. Thesis. University of Warwick.

Ostlund, S., Pandit, R., Rand, D. A. & Siggia, E. 1982 The 1-dimensional Shrödinger equation with an almost-periodic potential. *Phys. Rev. Lett.* **50**, 1873.

Ostlund, S., Rand, D. A., Siggia, E. & Sethna, J. 1983 Universal properties of the transition from quasi-periodicity to chaos. *Physica* D, **8**, 303–342.

Rand, D. A. 1984 Universality for golden critical circle maps and the breakdown of golden invariant tori. Cornell Preprint. (Submitted to *Comm. Math. Phys.*)

Rand, D. A. 1986 The singularity spectrum for hyperbolic Cantor sets and attractors. Preprint. University of Warwick.

Rand, D. A. 1987 Universality for the breakdown of dissipative golden invariant tori. In *Proc. 8th International Congress of Mathematical Physics, Marseilles, 1986* (ed. M. Mebkhout & R. Sénéor), pp. 537–547. World Scientific.

Satija, I. I. 1987 Universal strange attractor underlying Hamiltonian stochasticity. *Phys. Rev. Lett.* (Submitted.)

Schmidt, G. & Bialek, J. 1982 Fractal diagrams for Hamiltonian stochasticity. *Physica* D **5**, 397–404.

Umberger, D., Farmer, J. D. & Satija, I. I. 1986 A universal attractor underlying the quasi-periodic transition to chaos. *Phys. Lett.* A **114**, 341.

Yoccoz, J.-C. 1984 Conjugaison differentiable des diffeomorphismes du cercle dont le nombre de rotation verifie une condition Diophantienne. *Ann. Scient. Éc. norm. sup., Paris* **17**, 333–361.

Discussion

L. GLASS (*Department of Physiology, McGill University, Montreal, Canada*). Inside the Arnol'd tongues there are also very delicate bifurcations. For continuous non-invertible one-dimensional circle maps, the bifurcations in the Arnol'd tongues can be studied by a consideration of the locus of superstable cycles (the 'skeleton'), (Glass & Perez 1982; Belair & Glass 1985). Does Professor Rand's theory allow one to derive the bifurcation structure inside the Arnol'd tongues?

References

Belair, J. & Glass, L. 1985 *Physica* D **16**, 143.
Glass, L. & Perez, R. 1982 *Phys. Rev. Lett.* **48**, 1772.

D. A. RAND. The specific renormalization strange set that I have discussed does not give any insight into the detailed period-doubling structure in the Arnol'd tongues except to relate the structure in a tongue to those obtained after renormalizing a number of times. However, the general formalism is likely to be of use in describing this in the following way. J. van Zeijts in his Ph.D. thesis with R. MacKay has analysed the detailed fractal structure of the various two-codimensional period-doubling accumulation points using a similar formalism based on two transformations Q and R. The basic idea is that at a two-codimensional point the way in which an orbit of a critical point visits the neighbourhood of the two critical points gives a well-defined sequence which determines the order in which to apply Q and R to renormalize the system. The possible choice will correspond to the set of one-sided infinite sequences of 0s and 1s. There should be a way of appending Q and R to the transformations T_n that I described so that they will be defined on subsets of the excluded regions $\rho = 1/n$. Then, as for the T_ns alone, all these transformations can be joined together to get a 'super renormalization' transformation T which will have a strange set describing both the universal structure of the transition I described and the complex universal structure associated with the period-doubling curves in each tongue. To renormalize about one of these latter points in the (p/q)-tongue one would apply the appropriate sequence of T_ns to get the rotation number of the form $1/n$ and then apply the sequence of Qs and Rs to get at the period-doubling structure.

Reference

van Zeijts, J. 1987 A universal Cantor set for a renormalization semigroup with two generators for maps of the interval. Ph.D. thesis, Twente University, Holland.

From chaos to turbulence in Bénard convection

By A. Libchaber

James Franck and Enrico Fermi Institutes, University of Chicago, Chicago, Illinois 60637, U.S.A.

Recent results showing multifractal dimensions of attractors at the transition to a chaotic state for small-aspect-ratio cells of mercury are presented. Also as the Rayleigh number is increased up to 10^{11} in He gas a transition from a chaotic state to a 'soft turbulent state' and for a higher value of the control parameter to a 'hard turbulent state' is proposed.

1. Introduction

Here I present some recent results on Bénard convection, limited to the understanding of the various transitions to chaos and turbulence. The experiments are restricted to fluids with low Prandtl number; mercury first and helium gas next. Also the geometry of the cell is such that the aspect ratio is small; thus above the onset of convection one or two rolls can be accommodated. As has been noted, this allows the time-dependent state to be analysed in terms of dynamical-systems theory with a small number of degrees of freedom, for low Rayleigh numbers. In the first part, results of a mercury experiment are presented with an analysis of the transition to chaos (Stavans *et al.* 1985; Glazier *et al.* 1986) and the multifractal aspect of the attractor at the critical point (Jensen *et al.* 1985).

In the second part, an experiment in He gas similar in scope to that of Threlfall (1975) is presented, where the beginning of a study of transitions from chaos to turbulence is undertaken. Here we postulate the existence of two turbulent states, related to different dynamical behaviour of the boundary layer. In this approach the suggestion is that the transition from a chaotic state to soft turbulence comes from the formation of a boundary layer that detaches at the corners of the cell, thus leading to space-dependent behaviour similar to that found in large-aspect-ratio cells. The second transition in this view will be associated with the formation of a turbulent boundary layer and emission of plumes. The chaotic state itself is in this view related only to dynamical behaviour of the rolls, without any boundary-layer dynamics.

2. A forced Bénard experiment in mercury

In this fluid-flow experiment we wished to address the question of quasi-periodicity. The practical approach chosen was to start from a state of the fluid above the onset of convection, where the time-dependent state is associated with the oscillatory instability (Busse 1978).

The experiment was performed in the context of Rayleigh–Bénard convection. In a Rayleigh–Bénard experiment, a horizontal layer of fluid enclosed by two horizontal plates is heated uniformly from below. The fluid's density at the bottom of the layer is then smaller than at the top. When the temperature gradient across the layer is high enough, convection sets in because of the unstable density gradient. A rescaling of the equations of motion shows that the problem is controlled by two non-dimensional numbers: the Rayleigh number R and the Prandtl number P. Whereas R is proportional to the temperature gradient and thus contains geometrical information about the system, P is an intrinsic property of the fluid. We used mercury ($P = 0.025$) as a working fluid. When convection sets in at $R = R_c$ it takes the form of horizontal rotating rolls whose lateral dimension is of the order of the layer thickness $d = 0.7$ cm. Adjacent rolls rotate in opposite directions. In our case the cell was 0.7 cm × 0.7 cm × 1.4 cm and two convective rolls were present.

When R is increased beyond R_c, the convective roll pattern eventually becomes unstable. For low-P fluids, like mercury, the system undergoes a Hopf bifurcation into a time-dependent periodic mode owing to the oscillatory instability (oi) (Busse 1978). The mode is characterized by an AC vertical vorticity otherwise absent in the static roll pattern. Its period is of the order of d^2/k where k is the thermal diffusivity of the fluid. The oscillation is one of our two oscillators. For our cell, the frequency of the oi, w_i, was typically 230 mHz.

The second oscillator was introduced electromagnetically because mercury is an electrical conductor. An AC electrical current sheet was passed through the mercury and the whole system was immersed in a horizontal magnetic field ($H = 2 \times 10^{-2}$ T) parallel to the roll axes. The geometry of the electrodes and field was such that the Lorentz force on the fluid produced vertical vorticity. In this way the two oscillators were dynamically coupled. The value of current used was typically 20 mA. During the experiment R was held fixed at $R = 4.09R_c$ so that the oi had a high enough amplitude. In our case, the critical temperature difference across the layer at which convection started was 3 K. The temperature stability of the plates in the experiment was controlled to 10^{-5} K. A change of one millidegree induced a change of 10 µHz in w_i. A signal was obtained from the experiment by means of a thermal probe located in the bottom plate of our cell.

The nonlinear interaction between both oscillators was controlled by the amplitude of the injected current. The introduction or change of the external excitation's amplitude induced a change in w_i and in the amplitude of the oi. The change in the latter was not appreciable as long as the initial amplitude was high enough. Owing to these changes, the external oscillator's frequency w_e was adjusted after every change in the current's amplitude to achieve a particular winding number $\sigma = w_i/w_e$, namely either the golden mean σ_g or the silver mean σ_s. The winding number was measured by observing the frequencies of both oscillators in a fast Fourier spectrum of the time signal, and by following locked states corresponding to the rational approximants of the winding number. Winding numbers were approximated to within 2×10^{-4}.

Let us consider two oscillators that are nonlinearly coupled and assume that initially the coupling is weak. Then two possible situations may arise: the winding number, i.e. the frequency ratio between both oscillators, can be either a rational or an irrational number. For the former, the signal from the experiment is periodic.

FIGURE 1. The experimental attractor in two dimensions. 2500 points are plotted. The winding number is close to the golden mean and the experiment is at the point of breakdown of a 2-torus i.e. the critical point.

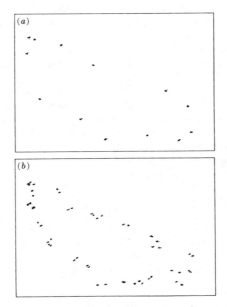

FIGURE 2. Attractor for an 8/13 tongue. (*a*) pure 8/13 state; (*b*) cascade of period-doubling up to period 8.

The oscillators are locked. For the latter the signal exhibits a beating pattern and the motion is quasiperiodic. In contrast to a linear coupling, the interaction shifts the frequencies of the oscillators so that locked states have finite width: when one of the frequencies changes within a finite range, the other frequency changes as well so that the winding number does not change. Defining a parameter space by plotting the amplitude of the interaction against the inverse frequency of one of the two oscillators, one can outline regions, called Arnol'd tongues, where the oscillators are locked. The width of these regions increase with the amplitude of the coupling. There is one tongue for each rational number.

Eventually the tongues overlap defining a line, the 'critical line'. The tongues are ordered through the Farey construction, namely: for any two tongues with winding numbers p/q and p^1/q^1 (parents), the tongue $(p+p^1)/(q+q^1)$ is found between them. This tongue is the one with the smallest denominator of all tongues between the two parents. The intersection of all tongues with the critical line forms a Cantor-like set characterized by a fractal dimension D.

At criticality one can plot the phase space attractor for a quasiperiodic state near the golden mean (figure 1) and for a period-doubling cascade on an 8/13 tongue, one of the Fibonacci approximants to the golden mean (figure 2).

3. Measurement of the spectrum of singularities at the transition points

To examine global scaling properties it is not sufficient to measure the dimension of the attracting set; the set certainly contains more topological information than can be characterized by a single number. To achieve a characterization that more fully describes those properties of such sets which remain unchanged under smooth changes of coordinates, it has been proposed to use a continuous spectrum of scaling indices. These spectra display the range of scaling indices and their density in the set. To clarify what we mean, consider the experimental cycle displayed in figure 1. It is clear that the time series is concentrated with various intensities in different regions. The spectrum that we use quantifies this variation in density on the attractor, and allows us to show the similarity of this cycle to sets produced by model equations that describe the onset of chaos via quasiperiodicity. In fact, the approach proposed here constitutes a rare opportunity for an extensive quantitative comparison of experiments with universal results obtained from theoretical models.

The theoretical work is aimed at estimating in a quantitative fashion how 'bunched' the density on the orbit might be. In technical terms this bunching is a description of singularities in the probabilities of the orbit points. Less technically, one can view a particular orbit point x_i, in a phase space like that of figure 1, and ask what is the probability of other points falling within a small distance, l, of this one. Call this probability $p_i(l)$. One can describe this probability by defining an index $\alpha_i(l)$ such that

$$P_i(l) = l^{\alpha_i(l)}. \qquad (1)$$

In typical sets the scaling index α_i takes, for small l, a range of values between α_{min} and α_{max}. We refer to this situation as a spectrum of singularities.

A function $f(\alpha)$ was introduced (Halsey *et al.* 1986) to measure for any α the probability density. The results for quasiperiodicity and period doubling are shown in figure 3. They define for the transition to chaos in those two general cases the global properties of the attractor. This problem illuminates the essential understanding of the routes to chaos up to the transition point. There, a complete agreement between theory and experiment is observed.

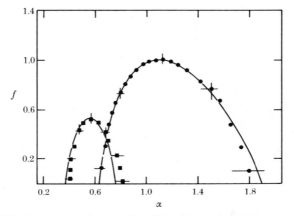

FIGURE 3. The $f(\alpha)$ spectrum for period-doubling (left curve) and quasiperiodicity at the golden mean (right curve). The dots show the experimental points.

As far as the chaotic state itself is concerned the main experimental tool is related to the dimension of the attractor, following the Grassberger & Procaccia (1983) algorithm. Unfortunately this technique works only for small dimensions. Essentially the number of points needed to measure it scales with a power of the dimension and soon becomes intractable. Nevertheless, the transition to chaos and the chaotic state can be experimentally defined. One of the experimental signatures of the chaotic state is that any local measurement should give the general picture of the dynamics. In other words, two measuring probes should show a coherence function between them close to unity.

An important aspect of the problem of turbulence then becomes the following. In a given experiment when does one observe a more complex dynamics where time and space become relevant? To answer such questions we have designed an experiment where a very high Rayleigh number, of the order 10^{12}, can be reached, whereas in the preceding ones the Rayleigh number could reach about 10^5 at most.

4. TRANSITIONS TO TURBULENCE IN He GAS

Let us just sketch first results for a Rayleigh–Bénard experiment in He gas at low temperature (4° K) and very high Rayleigh numbers (10^{12}), for a cylindrical cell of height and diameter 8.7 cm. As shown by Threlfall (1975) it is possible to cover 11 orders of magnitude in the Rayleigh number, with very small changes in the Prandtl number, by changing the pressure inside the cell. Careful measurements of the Nusselt number (relative effective conductivity of the gas) led Threlfall to discover several transitions in the buoyancy state. In the experiment presented here, our main emphasis is on local measurements of the temperature at two different points inside the cell to characterize the various transitions. The main results are as follows. At low Rayleigh number we observe the onset of the oscillatory instability followed by the now well-known routes to chaos. We label the chaotic state itself in terms of the attractor's correlation dimension. We then observe the transition to spatial disorder through the loss of coherence between detectors for $Ra \approx 2.5 \times 10^5$. At higher Rayleigh numbers two successive régimes with slightly different power laws for the Nusselt number are distinguished; the transition in our geometry begins at $Ra = 4 \times 10^7$. The time recordings, power spectra and histograms of the temperature fluctuations are different for the two régimes. A possible but not corroborated explanation for the two régimes is associated with a laminar and a turbulent boundary layer. The Nusselt against Rayleigh number curve is shown in figure 4, with the various observed dynamical régimes.

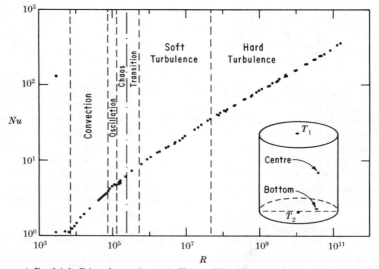

FIGURE 4. Rayleigh–Bénard experiment in He gas. Plot of the Nusselt number against Rayleigh number. The various dynamical states are labelled. The insert shows the cell and position of bolometers used to characterize the states.

The soft turbulence régime is associated with detachment of a laminar boundary layer, leading to the presence of many rolls in the cell, and thus to large-aspect-ratio behaviour. The hard turbulence state is related to a turbulent boundary layer with bursts of thermal plumes. This experiment is now in progress (Heslot et al. 1987).

References

Busse, F. H. 1978 *Rep. Prog. Phys.* **41**, 1929.
Grassberger, P. & Procaccia, I. 1983 *Physics Lett.* A **50**, 346.
Glazier, J. A., Jensen, M. H., Libchaber, A. & Stavans, J. 1986 *Phys. Rev.* A **34**, 1621.
Halsey, T. C., Jensen, M. H., Kadanoff, L. P., Procaccia, I. & Shraiman, B. 1986 *Phys. Rev.* A **33**, 1141.
Heslot, F., Castaing, B. & Libchaber, A. 1987 Transitions to turbulence in He gas. Preprint.
Jensen, M. H., Kadanoff, L. P., Libchaber, A., Procaccia, I. & Stavans, J. 1985 *Phys. Rev. Lett.* **55**, 2798.
Stavans, J., Heslot, F. & Libchaber, A. 1985 *Phys. Rev. Lett.* **55**, 596.
Threlfall, D. C. 1975 *J. Fluid Mech.* **67**, 17.

Dynamics of convection

By N. O. Weiss

*Department of Applied Mathematics and Theoretical Physics,
University of Cambridge, Cambridge CB3 9EW, U.K.*

Thermal convection in a fluid layer is an example of a dynamical system governed by partial differential equations. As the relevant control parameter (the Rayleigh number) is increased, successive bifurcations may lead to chaos and the nature of the transition depends on the spatial structure of the flow. Numerical experiments with idealized symmetry and boundary conditions make it possible to explore nonlinear behaviour in some detail and to relate bifurcation structures to those found in appropriate low-order systems. Two examples are used to illustrate transitions to chaos. In two-dimensional thermosolutal convection, where the spatial structure is essentially trivial, chaos is caused by a heteroclinic bifurcation involving a symmetric pair of saddle foci. When convection is driven by internal heating several competing spatial structures are involved and the transition to chaos is more complicated in both two- and three-dimensional configurations. Although the first few bifurcations can be isolated a statistical treatment is needed for behaviour at high Rayleigh numbers.

1. Spatial structure and temporal chaos

Most theoretical studies of temporal chaos in dissipative systems have been concerned with low-order systems whose behaviour can be analysed in detail. These simple models can be related to laboratory experiments or to numerical solutions of partial differential equations, where spatial structure is significant. Fluid dynamics provides a range of problems in which complicated spatio-temporal behaviour has been observed: thermal convection and flow between differentially rotating cylinders are classical examples (Guckenheimer 1986). In Taylor–Couette flow certain bifurcations involving changes in spatial structure and transitions to time-dependent and chaotic motion have been observed both in laboratory experiments and in precise numerical simulations (see, for example, Cliffe & Mullin 1985; Mullin *et al.* 1987). Laboratory experiments on convection show periodic, quasiperiodic and chaotic oscillations (see, for example, Libchaber *et al.* 1983) and the instabilities of two-dimensional rolls have been exhaustively investigated (Busse 1981). In what follows I shall discuss idealized numerical experiments that illustrate the development of temporal chaos in convecting systems with essentially trivial spatial structure and can be extended to explore the effect of introducing non-trivial spatial structure.

In any chaotic system it is important not only to establish the route to chaos but also to isolate the mechanism that causes its appearance. In many examples chaos originates from homoclinic or heteroclinic bifurcations, when a limit cycle or a torus collides with a non-stable singular point. To explain the origin of chaos

[71]

in nonlinear convection it is often necessary to represent the relevant partial differential equations by low-order model systems that are amenable to analysis. These toy systems may be constructed intuitively or derived in some asymptotic limit but their significance transcends the particular circumstances of their conception.

Much of our understanding of dissipative chaos has been gained by studying toy problems. In the early 1960s chaotic behaviour was recognized in several third-order autonomous systems, including models of coupled disc dynamos (Allan 1962) and thermosolutal convection (Moore & Spiegel 1966). At the same time Saltzman and Veronis realized that attempts to solve the partial differential equations governing Rayleigh–Bénard convection with severely truncated Galerkin expansions often generated spurious aperiodicity. The great achievement of Lorenz (1963) was to appreciate the significance of the lowest-order model. The Lorenz equations can be written in the form

$$\dot{a} = \sigma(-a+rb), \quad \dot{b} = -b+a(1-c), \quad \dot{c} = \nu(-c+ab), \qquad (1)$$

where r, σ and ν are real positive parameters; r is proportional to the Rayleigh number, which measures the rate of heating from below, and σ (the ratio of viscous to thermal diffusivity) and ν (a geometrical factor) are conventionally set to 10 and $\frac{8}{3}$, respectively. Lorenz first analysed this system and then used it to demonstrate sensitive dependence on initial conditions. The trivial solution $a = b = c = 0$ loses stability in a supercritical pitchfork bifurcation at $r = 1$. The two solution branches that emerge correspond to steady convection and remain stable up to $r \approx 24.74$, when they undergo subcritical Hopf bifurcations, shedding unstable limit cycles which are destroyed in homoclinic bifurcations at $r \approx 13.93$. It is this homoclinic explosion that ultimately leads to the appearance of a strange attractor (Sparrow 1982).

There are physical systems that are modelled faithfully by the Lorenz equations but two-dimensional Rayleigh–Bénard convection is not one of them. Reliable numerical solutions of the partial differential equations yield the bifurcation pattern sketched in figure 1a. The Lorenz system (1) is constructed to describe the initial pitchfork but in the full system the Hopf bifurcation (at $r \approx 90$) has become supercritical and there is no sign of chaos (Moore & Weiss 1973; Curry et al. 1984). To be sure, chaotic behaviour has been found in other configurations with similar bifurcation patterns. When convection is driven by internal heating complicated behaviour results from instabilities in the thermal boundary layer, as we shall see in §3. Chaos appears also for two-dimensional convection in a porous medium (Kimura et al. 1986) and can scarcely be avoided if three-dimensional perturbations are admitted (Curry et al. 1984). Nevertheless, toy systems must be used with caution: it is not always obvious that the full equations will repeat the bifurcation pattern of a truncated model.

There are, however, fluid systems that behave qualitatively like simple models. Here the partial differential equations can be represented by a low-order system of ordinary differential equations. It can be proved that solutions are accurately described by a large but finite number of modes (Foias et al. 1983). In such systems these modes are slaved to a few order parameters. In other words the dynamics

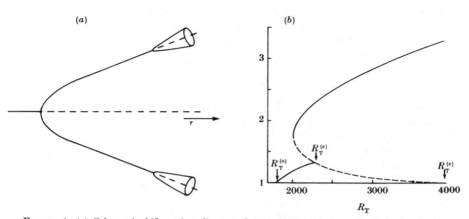

FIGURE 1. (a) Schematic bifurcation diagram for two-dimensional convection, showing a measure of the amplitude plotted against the stability parameter r. The initial pitchfork bifurcation is followed by a pair of secondary Hopf bifurcations. Unstable solutions are denoted by broken lines. (Based on calculations by Moore & Weiss (1973).) (b) Bifurcation diagram for thermosolutal convection, showing a quadratic measure of convective amplitude (the thermal Nusselt number N_T) plotted against the thermal Rayleigh number R_T. The branch of oscillatory solutions terminates in a heteroclinic bifurcation when $R_T = R_T^{(c)}$. (Based on calculations by Huppert & Moore (1976).)

is described by motion on a low-dimensional 'inertial manifold'. In some chemical reactions there is a close correspondence between experimental results and the behaviour of low-order systems (Argoul *et al.* 1987*a,b*; Richetti *et al.* 1987). In other cases (e.g. triple convection) the model can be systematically derived as normal form equations (Arnéodo *et al.* 1985*a*; Arnéodo & Thual 1985, 1987). In the next section I describe an example where the appearance of chaos in a heteroclinic bifurcation is accurately represented by a third-order toy system. Then, in §3, I discuss the complications that arise when several different spatial structures are involved. Finally, I speculate on the behaviour of systems with many competing spatial modes.

2. CHAOS IN DOUBLE-DIFFUSIVE CONVECTION

One way of generating self-excited oscillations is to modify a lightly-damped oscillator by adding a small destabilizing term that is out of phase with the restoring force. The linear system

$$\ddot{x} + 2\lambda\dot{x} + (1 - 2i\mu) x = 0, \tag{2}$$

where λ and μ are small positive constants, has solutions of the form

$$x \approx \exp{(\mu - \lambda)} t \exp{i[\{1 + \tfrac{1}{2}(\mu^2 - \lambda^2)\} t]}, \tag{3}$$

which grow exponentially for $\mu > \lambda$. (This is the mechanism put forward by Rayleigh (1896) to explain the Rijke tube phenomenon and taken up in

Eddington's (1926) theory of pulsating stars.) The amplitude of the oscillations is then limited by nonlinear effects: if the control parameter μ is increased there may be transition to chaos. For example, the system

$$\left.\begin{array}{l} \ddot{x}+2\dot{x}+(1-2\mathrm{i}\mu)\,x = \mathrm{i}\mu\bar{x}y, \\[2mm] \dot{y}+\nu y = -\mathrm{i}x(\dot{x}+x)/2\mu \quad (0<\nu<1), \end{array}\right\} \tag{4}$$

which is a sixth-order complex generalization of the Lorenz equations, shows periodic and quasiperiodic behaviour, followed by frequency locking and a period-doubling cascade that leads to chaos (Jones *et al.* 1985).

A fluid dynamical analogue of this process can be constructed as follows. Consider a layer of fluid with a dissolved solute (e.g. salty water) confined between two horizontal planes with boundary conditions such that the fluid is hot and salty at the lower plane but cold and fresh at the top. If the solute gradient is large enough the layer will be stably stratified and a fluid element will oscillate about its mean position. Now the ratio, τ, of the solutal diffusivity to the thermal diffusivity is small and this leads to what Eddington called overstability. As the fluid element rises it retains its salt content but loses heat to its surroundings; hence it is denser when it passes its original position and overshoots on the far side, where it gains heat and so goes further on the return trip (Stern 1960).

Two-dimensional thermosolutal convection with idealized boundary conditions has been studied in considerable detail (Veronis 1968; Huppert & Moore 1976; Moore *et al.* 1983; Knobloch *et al.* 1986). The relevant parameters are the thermal Rayleigh number R_T and the solutal Rayleigh number R_S (which measure the destabilizing temperature gradient and the stabilizing gradient of solute concentration respectively) together with the diffusivity ratio τ and the aspect ratio Λ. The system possesses a trivial static (conducting) solution for all parameter values, which becomes unstable as R_T is increased. For $\tau < 1$ and R_S sufficiently large there is a Hopf bifurcation when $R_T = R_T^{(o)}$ and convection sets in as overstable oscillations. If R_T is further increased, there is a stationary bifurcation at $R_T = R_T^{(e)}$. By varying R_S one can locate the degenerate bifurcation of codimension two, when $R_T^{(o)} = R_T^{(e)}$; the corresponding normal form equations are those for a Bogdanov bifurcation (Arnol'd 1983; Guckenheimer & Holmes 1983). Weakly nonlinear theory can thus be used to show that for $R_T^{(e)} - R_T^{(o)}$ sufficiently small the branch of oscillatory solutions terminates on the unstable steady branch in a heteroclinic bifurcation at $R_T = R_T^{(c)}$, where $R_T^{(o)} < R_T^{(c)} < R_T^{(e)}$ (Knobloch & Proctor 1981). Phase portraits when R_T is just less than, equal to and just greater than $R_T^{(c)}$ are sketched in figure 2a. The symmetrical pair of (non-stable) saddle points corresponds to the unstable steady solutions (with clockwise and anticlockwise motion). The limit cycle swells until it collides with the saddle points and is destroyed; at $R_T = R_T^{(c)}$ there is a heteroclinic orbit with infinite period and thereafter all trajectories escape from the neighbourhood of the origin.

The same pattern is revealed in numerical experiments. Figure 1b shows a bifurcation diagram in which the Nusselt number N_T (a quadratic measure of the amplitude of convection) is plotted against R_T. Once again the oscillatory branch terminates in a heteroclinic bifurcation but trajectories are attracted to the upper,

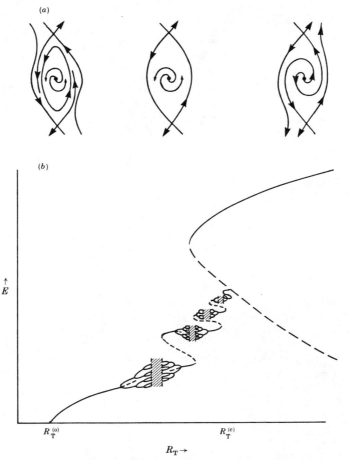

FIGURE 2. (a) Phase portraits in the neighbourhood of a heteroclinic bifurcation, showing a limit cycle for $R_T < R_T^{(c)}$, a heteroclinic orbit for $R_T = R_T^{(c)}$ and almost all trajectories escaping from the neighbourhood of the origin for $R_T > R_T^{(c)}$. (b) Schematic bifurcation diagram for thermosolutal convection with chaotic oscillations, showing the kinetic energy E as a function of R_T. Regions of chaos appear within bubbles enclosed by cascades of period-doubling bifurcations. (After Knobloch *et al* 1986.)

stable portion of the steady branch. For higher values of R_S solutions behave in a more complicated way. The schematic bifurcation diagram in figure 2b, in which the kinetic energy E is plotted against R_T, shows a sequence of 'bubbles' separated by saddle-node bifurcations. In each bubble there is an interval of chaos enclosed by two cascades of period-doubling bifurcations. Figure 3 shows projections of trajectories onto the plane with coordinates u and $\dot{u}$, where u is the horizontal

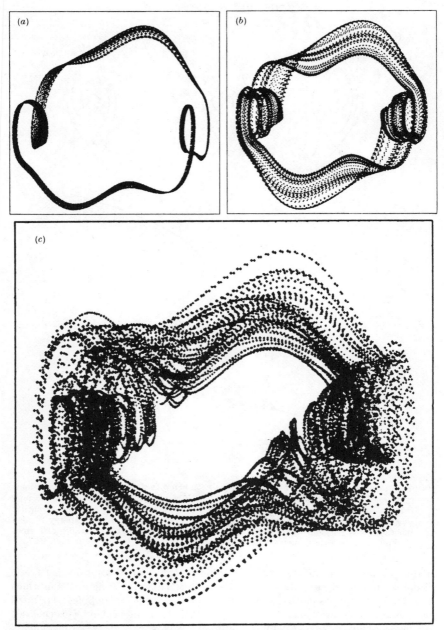

FIGURE 3. Portraits of chaos in thermosolutal convection. Projections of trajectories onto the u–$\dot{u}$ plane. (a) Semiperiodic behaviour for $R_{\mathrm{T}} = 10475$, (b) almost symmetrical chaos for $R_{\mathrm{T}} = 10508$ and (c) fully developed chaos for $R_{\mathrm{T}} = 11000$, close to the heteroclinic bifurcation. (After Knobloch *et al.* 1986.)

velocity half way across the base of the cell. The chaotic attractor is depicted by plotting a point at each timestep of the calculation. Semiperiodic behaviour is shown in figure 3*a*, where trajectories wander within a tube enclosing the (asymmetrical) unstable periodic orbit. As R_T is increased the attractor develops more structure, as shown in figure 3*b*. Figure 3*c* shows the chaotic attractor for a value of R_T that is close to $R_T^{(c)}$. This is a nice image of chaos but it is also instructive: trajectories spiral around the non-stable singular points and that provides a clue to the origin of chaos in this problem.

This mechanism can be established by studying toy problems. The partial differential equations can be reduced to a fifth-order system which is the analogue of the Lorenz equations for ordinary convection, except that it shares the bifurcation structure of the full system (Veronis 1965; Da Costa *et al.* 1981). In the limit $\tau \to 0$ the fifth-order model simplifies to the third-order system

$$\dot{a} = \tilde{r}a - \tilde{s}b, \quad \dot{b} = -b + a(1-c), \quad \dot{c} = \nu(-c+ab), \tag{5}$$

where $\tilde{r}$, $\tilde{s}$ are related to the thermal and solutal Rayleigh numbers (Knobloch *et al.* 1987). The substitution $\tilde{r} = -\sigma$, $\tilde{s} = -\sigma r$ reduces (5) to (3), so this system is equivalent to the Lorenz equations with σ negative. (This parameter range is also relevant in laser physics (Elgin & Molina Garza 1987).) The toy system (5), like the third-order model of Moore & Spiegel (1966), exhibits chaotic behaviour near the end of the oscillatory branch. The transition to chaos is related to the eigenvalues on the unstable steady branch. This branch bifurcates from the trivial solution at $\tilde{r} = \tilde{s}$, where there are three real eigenvalues λ_i such that $\lambda_1 < 0$, $\lambda_2 = 0$ and $\lambda_3 > 0$. As $|a|$ increases λ_2 decreases until it merges with λ_3 to form a complex conjugate pair with negative real part. At the heteroclinic bifurcation, where $\tilde{r} = \tilde{r}^{(c)}$, trajectories therefore spiral into a symmetrical pair of saddle foci. Figure 4*a* shows an orbit near the end of the oscillatory branch, computed by J. B. Molina Garza. Now Shil'nikov (1965) proved that if $\lambda_3 > -\mathrm{Re}\,\lambda_1 > 0$ at a similar homoclinic bifurcation then the return map contains an infinite number of horseshoes, each of which is associated with infinite numbers of (non-stable) periodic and aperiodic orbits. This is the mechanism responsible for chaos in double-diffusive convection.

The approach to chaos has been clarified by recent work (Glendinning & Sparrow 1984; Gaspard *et al.* 1984; Arnéodo *et al.* 1985). As $\tilde{r}$ approaches $\tilde{r}^{(c)}$ the branch of oscillatory solutions wiggles towards heteroclinicity, as shown schematically in figure 4*b*, where the period P is plotted against $\tilde{r}$. On each successive wiggle the trajectory winds once more round the saddle foci, with a corresponding increase in the period. Moreover, the symmetrical periodic solutions undergo symmetry-breaking bifurcations, followed by period-doubling and chaos within bubbles terminated by inverse cascades. This structure explains the bubbles shown in figure 2*b* for the partial differential equations. Furthermore, the subsidiary solution branches formed in these bifurcations may themselves undergo homoclinic bifurcations, producing a more complicated structure. This is illustrated in figure 4*c*, which shows a number of different solution branches approaching homoclinicity for a fourth-order model of magnetoconvection (Bernoff 1987). Such intricate self-similar patterns can lead to a great variety of chaotic oscillations.

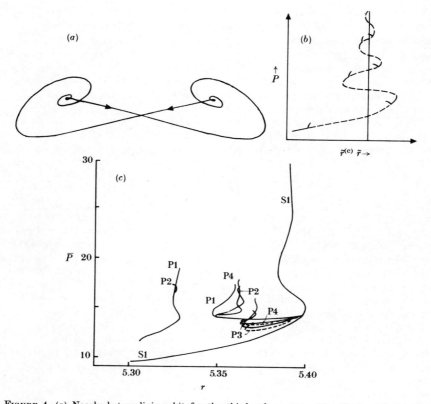

FIGURE 4. (a) Nearly heteroclinic orbit for the third-order system (5) with $\tilde{r} \approx 3.3$, $\tilde{s} = 53$, $\nu = 0.5$, projected onto the ac-plane. (After Elgin & Molina Garza 1987.) (b) Sketch showing period P as a function of $\tilde{r}$ in the neighbourhood of a heteroclinic bifurcation when Shil'nikov's condition is satisfied. (After Glendinning & Sparrow 1984.) (c) Multiple heteroclinic and homoclinic bifurcations for a fourth-order model of magnetoconvection. The normalized period $\bar{P}$ is plotted against the stability parameter r. The branch of symmetrical periodic (S1) solutions approaches a heteroclinic bifurcation, as in (b). The branches of asymmetrical (P1) solutions bifurcate from the S1 branch on either side of the first bubble and approach homoclinic bifurcations. Solutions of period two and four (P2, P4) appear at period-doubling bifurcations, whereas those of period three (P3, indicated by broken lines) are formed at saddle-node bifurcations; the corresponding solution branches apparently wiggle towards homoclinicity. (After Bernoff 1987.)

The thermosolutal problem provides the simplest illustration of a transition to chaos in double-diffusive convection. If the whole configuration is rotated the problem becomes more complicated, with a bifurcation of codimension three. It is then possible to derive normal form equations which allow chaotic behaviour arbitrarily close to the degenerate bifurcation (Arnéodo et al. 1985) and to compare results for this third-order system with those obtained from the partial differential equations (Arnéodo & Thual 1985, 1987). In a different context, experimental studies of chaos for the Belousov–Zhabotinsky reaction in stirred tank reactors have been related to homoclinic bifurcations in low-order model systems (Argoul *et al.* 1987*b*; Richetti *et al.* 1987).

These examples show that it is possible to set up fluid dynamical problems where the transition to chaos can be described by a toy system. This approach only succeeds if the continuous system is constrained by stringent symmetry or boundary conditions. As these conditions are relaxed behaviour grows more complicated. For example, relaxing the lateral boundary conditions in thermosolutal convection immediately allows travelling wave solutions (as well as the standing waves discussed above) resembling behaviour found in experiments on binary convection. Behaviour near the initial Hopf bifurcation can still be modelled by a toy system but its order is increased from two to four (Knobloch *et al.* 1986; Knobloch 1986). Increasing the aspect ratio in two-dimensional problems introduces an even greater variety of competing spatial structures, while three-dimensional behaviour is notoriously difficult to analyse.

3. COMPETING SPATIAL STRUCTURES

The complications introduced by systems with non-trivial spatial structure can be illustrated by examining a different configuration. Consider a fluid layer with a thermally insulating lower boundary, heated entirely from within (e.g. by radioactive decay). The appropriate Rayleigh number R is proportional to the volumetric heating rate. As R is increased the static solution loses stability in a pitchfork bifurcation (Roberts 1967). For R sufficiently large there is a thermal boundary layer at the upper boundary and motion is dominated by cold sinking plumes. As R is increased this boundary layer becomes unstable and there is a preference for narrower convection cells which leads to interesting dynamical behaviour.

Numerical experiments have concentrated on fluids with very high viscosity (motivated by convection in the earth's mantle). The bifurcation pattern for two-dimensional convection is identical to that in figure 1*a*. As R is increased there is a secondary Hopf bifurcation leading to oscillations about a steadily convecting state (Lennie *et al.* 1987). During each oscillation a cold blob develops in the thermal boundary layer and is swept into the descending plume. As it fades away a new blob develops at the top of the cell, so that there are two cold blobs at any instant, circulating as a modulated thermal wave (cf. Rand 1982). Further increases in R lead eventually to chaos.

The nature of the transition to chaos depends on the value of the aspect ratio Λ. When $\Lambda = 1.5$ there is an episode of chaos enclosed between two period-doubling cascades, recalling the example discussed in the previous section. The sequence of bubbles is, however, interrupted by a variety of periodic and quasiperiodic oscillations. Figure 5 shows isotherms at four equally spaced intervals during one period of an oscillation in this range. This solution clearly has non-trivial spatial structure, for the cold blob breaks away from the unstable layer before reaching the descending plume at the side of the cell. Dynamical constraints nevertheless

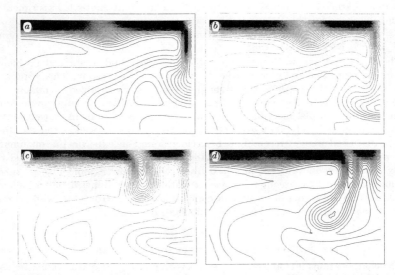

FIGURE 5. Periodic behaviour for two-dimensional convection driven by internal heating. Isotherms at equally spaced intervals during an oscillation for $R = 8 \times 10^5$, $\Lambda = 1.5$. Instability of the thermal boundary layer leads to non-trivial spatial structure. (After Lennie *et al.* 1987.)

ensure that the solution remains strictly periodic despite large deviations from the mean temperature field. The spatial structure in figure 5 involves modes that correspond to rolls of several different widths and the bifurcation pattern is consequently altered so that it deviates from that associated with Shil'nikov's mechanism in a third-order system.

As R increases the cold blob develops more rapidly and plunges downwards closer to the middle of the cell. Eventually the cell is split and the new plume migrates to the left-hand boundary (assuming that the original sense of motion was as shown in figure 5) where it is stabilized by the lateral boundary conditions. Meanwhile instabilities develop chaotically within the thermal boundary layer. Any description of this process has to take account of spatial structure. The primary transition is from a single roll solution, with a sinking plume at one boundary only, to a double roll solution, with plumes at both sides. The initial competition between these two solutions can be modelled by a second-order system that allows mixed mode solutions (Knobloch & Guckenheimer 1983). The original Hopf bifurcation could also be described within this framework and the phase portraits sketched in figure 6 show how the oscillations might develop. Figure 6a corresponds to the bifurcation diagram in figure 1a though only one solution branch is shown. In figure 6b the limit cycle has collided with a saddle point, corresponding to a non-stable mixed solution, and in figure 6c all

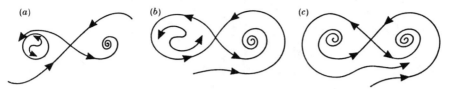

FIGURE 6. Homoclinic bifurcations in systems with competing spatial structures. Schematic phase portraits for a second-order model: (*a*) oscillations about a single roll solution, stable two-roll solution; (*b*) homoclinic bifurcation; (*c*) unstable single roll, stable two-roll solution.

trajectories are attracted to the stable (two-roll) solution. A higher-order extension of this system could then describe chaotic behaviour associated with the homoclinic bifurcation in figure 6*b*.

When *R* is sufficiently large other solutions, with three or more rolls, affect the dynamics of convection. Secondary and tertiary bifurcations then lead to families of mixed mode solutions and the periodic solution in figure 6 presumably belongs to such a family. The resulting dynamics could in principle be described by some system of relatively low order, which might still be regarded as a sophisticated toy. Experience with simpler problems (Nagata *et al.* 1987) suggests that it may no longer be instructive to work out the full bifurcation structure in such systems. It seems plausible that there will be many basins of attraction separated by non-stable fixed points in the multidimensional phase space and that there will be ample opportunity for chaos associated with homoclinic or heteroclinic bifurcations.

Numerical experiments on three-dimensional convection driven by internal heating provide a clearer impression of the competing planforms (McKenzie *et al.* 1987). Once again the velocity is determined by the temperature field, which is dominated by sinking plumes. Changes in the disposition of these plumes occur more readily when the fluid flow is no longer constrained to be two-dimensional. Moreover, these changes can be followed by studying the variation of temperature across a horizontal plane equidistant from the upper and lower boundaries. Stationary bifurcations may lead to changes of scale, such that the spacing between adjacent plumes is reduced as *R* increases, and to losses of symmetry in the convection pattern (McKenzie 1987). In addition there are oscillatory bifurcations and transitions to chaos with rich spatial structure. Figure 7 shows horizontal slices through the temperature field at six consecutive stages of a run. The sinking plumes can easily be recognized and the initial state in figure 7*a* deviates slightly from a symmetric solution with tetrad axes at each plume. There are two diagonal mirror planes with a dyad axis at the centre. The system loses symmetries as it evolves: by figure 7*c* only one mirror plane survives and that symmetry is broken in figure 7*e*. Thereafter the pattern varies chaotically though it is still possible to recognize ghost attractors with high symmetry.

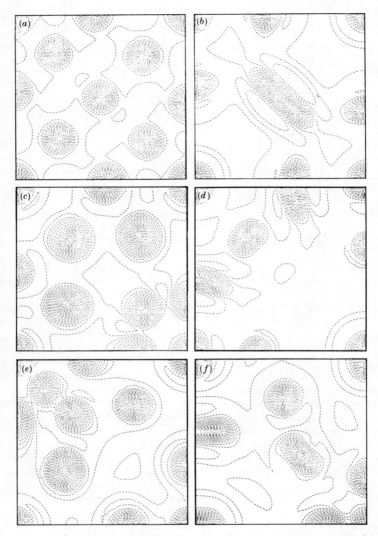

FIGURE 7. Aperiodic time-dependent behaviour for three-dimensional convection driven by internal heating. Isotherms in a horizontal plane through the middle of the layer, showing the location of cold sinking plumes, at equal intervals of time. The solution loses symmetries as time proceeds. (From computations by McKenzie *et al.* 1987.)

4. LARGE-ASPECT-RATIO SYSTEMS

The interaction between spatial and temporal behaviour can be explored in systems where only a limited number of spatial modes are involved. It is then possible to isolate successive bifurcations and to interpret the bifurcation structure in terms of some low-order model. Toy systems can thus be used to explain transitions to chaos and behaviour that is moderately complicated.

This approach has succeeded with systems that are constrained by stringent symmetry and boundary conditions, as in laboratory experiments with two or three convection rolls (see, for example, Bergé & Dubois 1979; Libchaber *et al.* 1983). The convective pattern grows more complicated as the aspect ratio is increased: experiments at high Rayleigh numbers show a combination of coherent structures and irregular aperiodic plumes (Busse 1981). Similar effects appear in numerical experiments. Time-dependence sets in at a lower Rayleigh number for higher aspect ratios and there is a richer variety of spatial structure when the preferred cell size is much smaller than the aspect ratio of the system.

So far, the effect of increasing the aspect ratio has been studied systematically only for one-dimensional models, which provide a link between toy systems and two- or three-dimensional behaviour. The best known example is the Kuramoto–Sivashinsky equation.

$$\partial_t u + \partial_x^2 u + \Lambda^{-2} \partial_x^4 u + u \partial_x u = 0, \tag{6}$$

with u constrained to be periodic in x. As the bifurcation parameter Λ is increased the trivial solution undergoes a sequence of bifurcations giving rise to multiroll solutions whose evolution can be followed (Hyman & Nicolaenko 1986; Frisch *et al.* 1986). For Λ sufficiently large solutions of (6) and other related systems exhibit spatio-temporal chaos, with complicated time-dependent modulation of the basic pattern (Chaté & Manneville 1987). It can nevertheless be shown that (6) is equivalent to a finite-dimensional system and that the dynamics occurs on an inertial manifold with a dimension that depends on Λ (Hyman & Nicolaenko 1986).

These models indicate a goal for theoretical studies of nonlinear convection. What is needed is to construct the relatively low-dimensional manifold on which the essential dynamics takes place, and to relate motion on that manifold to the evolution of coherent structures. Then we could discover how statistical properties of convection (such as the dimension of the attractor) vary as the Rayleigh number is increased.

Research described here has been supported by grants from SERC and I have benefited from numerous discussions with my colleagues.

REFERENCES

Allan, D. W. 1962 *Proc. Camb. phil. Soc.* **58**, 671–693.
Argoul, F., Arnéodo, A. & Richetti, P. 1987*a* *Phys. Lett.* A **120**, 269–275.
Argoul, F., Arnéodo, A., Richetti, P. & Roux, J. C. 1987*b* *J. chem. Phys.* **86**, 3325–3338.
Arnéodo, A., Coullet, P. H. & Spiegel, E. A. 1985*a* *Geophys. astrophys. Fluid Dyn.* **31**, 1–48.

84 N. O. Weiss

Arnéodo, A., Coullet, P. H., Spiegel, E. A. & Tresser, C. 1985 *Physica* D **14**, 327–347.
Arnéodo, A. & Thual, O. 1985 *Phys. Lett.* A **109**, 367–373.
Arnéodo, A. & Thual, O. 1987 *Springer series in synergetics.* (In the press.)
Arnol'd, V. I. 1983 *Geometrical methods in the theory of ordinary differential equations.* Berlin: Springer.
Bergé, P. & Dubois, M. 1979 *J. Phys. Lett.* **40**, L505–L509.
Bernoff, A. J. 1987 *Physica* D (Submitted.)
Busse, F. H. 1981 In *Hydrodynamics and the transition to turbulence* (ed. H. L. Swinney & J. P. Gollub), pp. 97–137. Berlin: Springer.
Chaté, H. & Manneville, P. 1987 *Phys. Rev. Lett.* **58**, 112–115.
Cliffe, K. A. & Mullin, T. 1985 *J. Fluid Mech.* **158**, 243–258.
Curry, J. H., Herring, J. A., Loncaric, J. & Orszag, S. A. 1984 *J. Fluid Mech.* **147**, 1–38.
Da Costa, L. N., Knobloch, E. & Weiss, N. O. 1981 *J. Fluid Mech.* **109**, 25–43.
Eddington, A. S. 1926 *The internal constitution of the stars.* Cambridge University Press.
Elgin, J. N. & Molina Garza, J. B. 1987 Preprint.
Foias, C., Manley, O. P., Temam, R. & Treve, Y. M. 1983 *Physica* D **9**, 157–188.
Frisch, U., She, Z. S. & Thual, O. 1986 *J. Fluid Mech.* **168**, 221–240.
Gaspard, P., Kapral, R. & Nicolis, G. 1984 *J. statist. Phys.* **35**, 697–727.
Glendinning, P. A. & Sparrow, C. T. 1984 *J. statist. Phys.* **35**, 645–696.
Guckenheimer, J. 1986 *A. Rev. Fluid Mech.* **18**, 15–31.
Guckenheimer, J. & Holmes, P. 1983 *Nonlinear oscillations, dynamical systems and bifurcations of vector fields.* New York: Springer.
Huppert, H. E. & Moore, D. R. 1976 *J. Fluid Mech.* **78**, 821–854.
Hyman, J. M. & Nicolaenko, B. 1986 *Physica* D **18**, 113–126.
Jones, C. A., Weiss, N. O. & Cattaneo, F. 1985 *Physica* D **14**, 161–176.
Kimura, S., Schubert, G. & Straus, J. M. 1986 *J. Fluid Mech.* **166**, 305–324.
Knobloch, E. 1986 *Phys. Rev.* A **34**, 1538–1549.
Knobloch, E., Deane, A. E., Toomre, J. & Moore, D. R. 1986 *Contemp. Math.* **56**, 203–216.
Knobloch, E. & Guckenheimer, J. 1983 *Phys. Rev.* A **27**, 408–417.
Knobloch, E., Moore, D. R., Toomre, J. & Weiss, N. O. 1986 *J. Fluid Mech.* **166**, 409–448.
Knobloch, E. & Proctor, M. R. E. 1981 *J. Fluid Mech.* **108**, 291–316.
Knobloch, E., Proctor, M. R. E. & Weiss, N. O. 1987 (In preparation.)
Lennie, T. B., McKenzie, D. P., Moore, D. R. & Weiss, N. O. 1987 *J. Fluid Mech.* (Submitted.)
Libchaber, A., Fauve, S. & Laroche, C. 1983 *Physica* D **7**, 73–84.
Lorenz, E. N. 1963 *J. atmos. Sci.* **20**, 130–141.
McKenzie, D. P. 1987 *J. Fluid Mech.* (Submitted.)
McKenzie, D. P., Moore, D. R., Weiss, N. O. & Wilkins, J. M. 1987 (In preparation.)
Moore, D. R., Toomre, J., Knobloch, E. & Weiss, N. O. 1983 *Nature, Lond.* **303**, 663–667.
Moore, D. R. & Weiss, N. O. 1973 *J. Fluid Mech.* **58**, 289–312.
Moore, D. W. & Spiegel, E. A. 1966 *Astrophys. J.* **143**, 871–887.
Mullin, T., Cliffe, K. A. & Pfister, G. 1987 Preprint.
Nagata, M., Proctor, M. R. E. & Weiss, N. O. 1987 (In preparation.)
Rand, D. 1982 *Arch. ration. Mech. Analysis* **79**, 1–37.
Rayleigh, Lord 1896 *The theory of sound*, 2nd edn, vol. 2. London: Macmillan.
Richetti, P., Roux, J. C., Argoul, F. & Arnéodo, A. 1987 *J. chem. Phys.* **86**, 3339–3356.
Roberts, P. H. 1967 *J. Fluid Mech.* **30**, 33–49.
Shil'nikov, L. P. 1965 *Sov. Math. Dokl.* **6**, 163–166.
Sparrow, C. T. 1982 *The Lorenz equations: bifurcations, chaos and strange attractors.* New York: Springer.
Stern, M. E. 1960 *Tellus*, **12**, 172–175.
Veronis, G. 1965 *J. mar. Res.* **23**, 1–17.
Veronis, G. 1968 *J. Fluid Mech.* **34**, 315–336.

Discussion

H. E. HUPPERT (*Department of Applied Mathematics and Theoretical Physics, University of Cambridge, U.K.*). Would Dr Weiss care to expand on his brief remarks on the new solutions which arise when the boundaries to a convection calculation are removed. Further, what would happen in the three-dimensional case, and how does he relate the results of his numerical experiments to the experimental results of Libchaber?

N. O. WEISS. The solutions are very sensitive to details of the boundary conditions. For instance, thermosolutal convection can set in as standing waves (or oscillations) with fixed lateral boundaries but as travelling waves if periodic boundary conditions are imposed. Moreover there can be transitions from travelling to standing waves involving branches of mixed (or modulated) solutions. Again, convection problems are affected when the aspect ratio is increased. As more cells are included behaviour becomes less regular and more difficult to analyse. In particular, oscillations and chaotic motion set in at lower Rayleigh numbers. This seems to be true both of numerical experiments and of laboratory experiments like those described by Libchaber.

Chaos: a mixed metaphor for turbulence

By E. A. Spiegel

Department of Astronomy, University of Columbia,
New York 10027, *U.S.A.*

There are special circumstances when the equations of fluid mechanics can be asymptotically reduced to third- or higher-order differential equations that admit chaotic solutions. For physically extended systems, similar reductions lead to simplified partial differential equations whose solutions contain coherent structures that interact in complicated and erratic ways.

It is suggested here that analogous reductions of the fluid equations are possible even when the fluid is in a turbulent state. From this we conclude that, more than being a metaphor for turbulence, chaos is a basic property of turbulent fluids.

1. Introduction

Many features of chaos are mirrored in turbulent flows and this has led some writers on chaos to use the words interchangeably. Both terms appear in ordinary language, where they do have an overlap of meaning, and both are used in scientific settings where they do not have generally accepted, precise definitions. Whether satisfactory definitions of these terms can be found before the subjects are better understood is not clear. But practitioners of both subjects can name the symptoms that should be present before a process can be considered to be chaotic or turbulent.

The property that a slight cause can produce a large effect is common to both subjects. It is understood that this property should be prevalent in maintained chaos or turbulence. Some students of chaos would consider this to be the defining property of chaos, given some further technical conditions. Few fluid dynamicists would accept this property as a complete characterization of turbulence. The conclusion is naturally that turbulence is a chaotic process, but it remains to be seen whether this notion will be of use. The advantage for fluid dynamicists of this viewpoint is that the theory of chaos makes processes like intermittency and the appearance of coherent structures mathematically accessible in certain restricted contexts. The existence of such partially solved problems to which aspects of turbulence may be reduced seems quite valuable.

What is important for fluid dynamics is that the theory of chaos continues to grow and to provide results that apply directly to fluids and that are also suggestive of turbulence. Studies of spatial chaos and spatio-temporal chaos are valuable in interpreting turbulent processes and advances in the theory of fractal sets have a bearing on some structural aspects of turbulence. If I incautiously write such things without yet giving a full list of turbulent symptoms, it is because fluid dynamicists use the word turbulence in diverse ways.

Real fluid turbulence is the nonlinear dynamics of a three-dimensional vorticity field; this is associated with the root *turbo*. But the root *turba* is the Latin word for mob, a 17th century British abbreviation for *mobile* and that may explain the use of the word turbulence to describe random wave fields (called undulence by some) or chaotic fields of thermals (thermalence). These various forms of turbulence have much in common: all may have vorticity and, I think, they are all chaotic. In what follows I shall offer fragmentary support of this remark.

2. CHAOS IN MACROSCOPIC SYSTEMS

On the mathematical side of the subject, chaos may be seen in discrete mappings of a space onto itself. Such mappings do arise in scientific problems, but the most natural description of the dynamics of extended systems, such as fluids, is in terms of differential equations. Chaos appears in solutions of ordinary differential equations of third and higher order, but these are not the form in which fluid problems are formulated either. Partial differential equations are the normal medium of expression of fluid problems.

Lorenz (1963) reduced the partial differential equations of two-dimensional, Boussinesq convection to a third-order system of ODEs by the use of a truncated expansion in normal modes. From this, he derived a map of a line into itself and thus clarified the arrival of chaos in differential equations. Because the chaos of the Lorenz system occurs in a range of fluid parameters where the original truncated expansion is quantitatively inadequate, some fluid dynamicists have concluded that the Lorenz equations do not apply to fluids.

Moore and I (Moore & Spiegel 1966) used qualitative physical arguments to devise another third-order ODE for the dynamics of a fluid parcel in doubly diffusive convection. This equation does lead to chaos, but fluid-dynamical colleagues could hardly credit the possibility of chaos in real fluids because of the loose way in which we derived the equations.

Despite the questionable provenances of these early descriptions, it is now known that these third-order systems of ODEs do apply to fluid dynamics in suitable circumstances, though perhaps not in the conditions under which they were first suggested. Moreover, though the conditions under which these equations are adequate asymptotic descriptions are specially selected, they may not be so special in turbulent fluids, as I shall suggest below. For these discussions it is useful to recall briefly the nonlinear theory of instability (see Stuart 1962).

The state of a fluid or other macroscopic system is governed by a set of fields $U_I(x, t)$, $I = 1, 2, \ldots, N$. These fields may be velocities, pressures, magnetic fields, and so on. We presume that the field equations are of the form

$$\partial_t \boldsymbol{U} = \boldsymbol{F}(U, \partial), \tag{2.1}$$

where ∂ is the spatial derivative and neither t nor ∂_t enter in $\boldsymbol{F}$ explicitly. Though (2·1) is not completely general (it does not serve for incompressible fluids) it will do for this discussion.

A stationary solution $\boldsymbol{U_0}$ of (2.1) satisfies

$$\boldsymbol{F}(\boldsymbol{U_0}, \partial) = \boldsymbol{0}. \tag{2.2}$$

Linear theory is described by the linearized form of (2.1)

$$\partial_t \boldsymbol{u} = \boldsymbol{M}\boldsymbol{u}, \tag{2.3}$$

where $\boldsymbol{M}$ is the operator

$$\boldsymbol{M} := \frac{\delta \boldsymbol{F}}{\delta \boldsymbol{U}}\bigg|_{\boldsymbol{U}=\boldsymbol{U}_0} \tag{2.4}$$

and δ indicates functional derivative. The linear problem has solutions like $\boldsymbol{\psi}(\boldsymbol{x})\exp(st)$.

Suppose that $\boldsymbol{F}$ depends on a parameter, R, and that we may find a range in R where $\mathrm{Re}\, s < 0$ for all but one of the linear solutions. For that exceptional solution, there is a neighbourhood of $R = R_0$ where

$$s = \mu(R), \quad \mu(R_0) = 0 \tag{2.5a,b}$$

and $\mu(R)$ is monotonic. In this neighbourhood, the solution of (2.1) is approximated as

$$\boldsymbol{U}(\boldsymbol{x}, t) = \boldsymbol{U}_0(\boldsymbol{x}) + A(t)\,\boldsymbol{\psi}(\boldsymbol{x}) + \text{corrections}, \tag{2.6}$$

where A is governed by the amplitude equation

$$\dot{A} = \mu A + g(A), \tag{2.7}$$

with $g(0) = 0$, $g'(0) = 0$. Various asymptotic methods are available to calculate the corrections to (2.6) and $g(A)$. A simple way to state the approach is through centre manifold theory (Carr 1981).

At $R = R_0$ the centre manifold theorem tells us that the time-dependence in the corrections of (2.6) can be expressed in terms of $A(t)$. The explicit calculation of the corrections can then be made in terms of what in fluid dynamics is called the Stuart–Watson method and in physics and nonlinear mechanics the Bogoliubov method. These come down to introducing the Ansatz (2.7), with

$$\boldsymbol{U}(\boldsymbol{x}, t) = \boldsymbol{U}_0(\boldsymbol{x}) + \boldsymbol{V}(\boldsymbol{x}, A(t)), \tag{2.8}$$

into (2.1) and developing in Taylor series in A (Coullet & Spiegel 1983).

It may happen that the instability occurs in the form of a growing oscillation. In that case, a pair of solutions, $\boldsymbol{\psi}(\boldsymbol{x})\exp(st)$ and its complex conjugate, are marginal at $R = R_0$. That is,

$$s = \pm \mathrm{i}\omega + \mu(R), \quad \mu(R_0) = 0. \tag{2.9a,b}$$

We can approximate the solution with

$$\boldsymbol{U}(\boldsymbol{x}, t) = \boldsymbol{U}_0(\boldsymbol{x}) + A(t)\,\boldsymbol{\psi}(\boldsymbol{x}) + A^*(t)\,\boldsymbol{\psi}^*(\boldsymbol{x}) + \text{corrections} \tag{2.10}$$

and

$$\dot{A} = (\mathrm{i}\omega + \mu)\,A + g(A, A^*). \tag{2.11}$$

Again, we can proceed to calculate the unknowns with the Ansatz

$$\boldsymbol{U}(\boldsymbol{x}, t) = \boldsymbol{U}_0(\boldsymbol{x}) + \boldsymbol{V}(\boldsymbol{x}, A, A^*) \tag{2.12}$$

together with (2.11).

Suppose now that another parameter, S, is placed at our disposal so that, in (2.9), instead of $\mu(R_0) = 0$ we have

$$\mu(R_0, S_0) = 0, \quad \omega(R_0, S_0) = 0. \tag{2.13}$$

In the neighbourhood of (R_0, S_0) in the parameter plane, we can have either a complex conjugate pair of modes, or a pair of monotonic modes like that of $(2.5a, b)$. This situation, with two parameters, was studied by Lyapounov and it arises whenever the frequency in an overstability may be tuned to zero. The modern discussion of this situation is that of Bogdanov. We now have a two-component amplitude A where the two components are independent, unlike the case of oscillatory instability. The coupled first-order equations for the two components of A may describe relaxation oscillation and limit cycles.

Suppose yet another parameter T is placed at our disposal. We may then be able to find a set (R_0, T_0, S_0) where a simultaneous monotonic and oscillatory bifurcation occur and where the frequency of the latter may be tuned to zero. This may be called a Lorenz bifurcation for, sufficiently close to (R_0, S_0, T_0) in parameter space and given suitable symmetry conditions, we shall asymptotically find equations like those studied by Lorenz.

The conclusion is that, if there are three tunable stability parameters, we may under suitable conditions reduce the field equations to ODEs that produce chaos for certain neighbourhoods of parameter space. A physical example is doubly diffusive convection with internal heat sources, a common situation in stellar interiors.

3. COHERENT STRUCTURES IN EXTENDED SYSTEMS

To get chaos in fluid dynamics by tuning three parameters at once, as discussed at the end of the last section, is possible, if academic (Arnéodo *et al.* 1982). However, if one of the parameters is the size of the system, it is possible to have many slightly unstable modes in the limit of large systems. Consider, for instance, fluid in a channel of depth unity and of infinite horizontal extent.

If we impose uniform forcing along the channel, the static solution U_0 of (2.2) will depend on z but not on x and y. So the solutions of (2.3) will be of the form

$$u(x, t) = \exp(st + ik_x x + ik_y y)\, \psi_n(z), \tag{3.1}$$

where s depends on $k = (k_x^2 + k_y^2)^{\frac{1}{2}}$, n and the parameter, R. The index $n = 1, 2, 3, \ldots$, is discrete but k ranges over a continuum of values. We assume that $s < 0$ for $n > 1$. For $n = 1$, there exists a pair of values, $k = k_c$ and $R = R_c$, for which $s = 0$, where $s < 0$ for $k \neq k_c$ and $R = R_c$. This is the simplest example of the onset of instability with a continuum of wavenumbers. In the neighbourhood of k_c and R_c, for $n = 1$, we have, approximately,

$$s = \alpha(R - R_c) + \beta(k^2 - k_c^2)^2, \tag{3.2}$$

where α and β are constants. This form ensures that $ds/dk = 0$ on $k = k_c$.

If we assume that only a very narrow band of wavenumbers controls the solution, we may proceed in the spirit of the Bogoliubov method with an Ansatz of the form (Coullet & Spiegel 1987)

$$U(x, t) = \psi(z) A(x, y, t) + \text{corrections}, \tag{3.3}$$

where

$$A(x, y, t) = \int A_k(t) e^{ik \cdot x} \, dk \tag{3.4}$$

and

$$\partial_t A_k = s(k) A_k + \Gamma_k [A_p]. \tag{3.5}$$

Here Γ_k is a strictly nonlinear functional of A_k. A widely studied model of an equation like (3.5) is the Swift–Hohenberg (1978) model of convection theory

$$\partial_t A = \alpha(R - R_c) A + \beta(\Delta + k_c^2)^2 A - A^3. \tag{3.6}$$

The actual nonlinear term for the case of convection is not so simple as this, except in the two-dimensional case. Then it may be reduced to the time-dependent Ginzburg–Landau equation, derived in fluid dynamics through asymptotic expansions. (For the case of convection theory, see Segel 1969; Newell & Whitehead 1969.)

As in the discrete case, when the instability is in the form of a growing oscillation, a complex pair of amplitude equations is found. The reduction to two dimensions leads to two equations for the amplitude of wave modulation, one each for waves going left and right. In the simplest case, with waves going in one direction only, we have a solution of the form

$$U(x, z, t) = e^{ik_c x - i\omega t} \psi(z) A(x, t) + \text{c.c.} + \text{corrections}, \tag{3.7}$$

where A satisfies the complex Ginzburg–Landau equation

$$i\partial_t A + \alpha A + \beta \partial_x^2 A + \gamma |A|^2 A = i\delta[\dots], \tag{3.8}$$

where α, β, γ, δ are real parameters. When $\delta = 0$, this is the cubic Schrödinger equation and that has a soliton solution

$$A(x, t) = \alpha S(x - Ut). \tag{3.9}$$

For δ small, but not zero, a reasonable limit in fluid dynamics, the situation is richer.

Numerical solutions (Moon *et al.* 1982) show that, as the computational domain is increased, temporal chaos erupts. Beyond that, the suggestion of spatio-temporal chaos emerges from numerical simulations showing a behaviour like that of a multisoliton gas (as shown in figure 1), except that the coherent structures in this chaos are not solitons; α and U are not constants but depend on δt (Bretherton & Spiegel 1983).

A further reduction is possible and has been achieved by several routes. In that of Kuramoto & Tsuzuki (1976), the development of (3.9) about a simple wave leads to an equation for the phase of the perturbation. This is the Kuramoto–Sivashinsky equation:

$$\partial_t \phi + \tfrac{1}{2}(\partial_x \phi)^2 = \text{dissipative terms}. \tag{3.10}$$

The dissipative terms here are linear highly differentiated terms that may contribute both positive and negative damping. A random looking field of shock-like

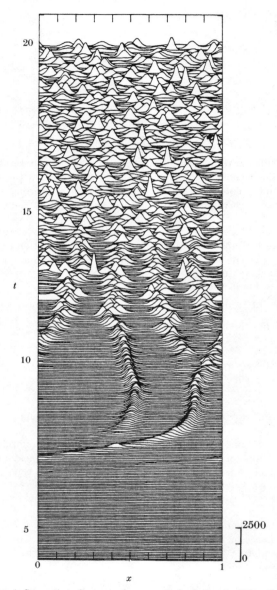

FIGURE 1. Space-time diagram of a numerical solution to (3.8), after
Bretherton & Spiegel (1983).

solitons is the issue. Whether this is strictly chaotic is a question clouded in technicalities that I will omit in favour of a more simple discussion.

If the solitary solutions of (3.10) may be approximated by discontinuities, that gives us some license to omit the dissipative terms in (3.10). With $u = \partial_x \phi$, we have the familiar

$$\partial_t u + u \partial_x u = 0. \tag{3.11}$$

Integrals of the form $\qquad u(x, t) = F(x - ut) \tag{3.12}$

give a hint of what happens in (3.11). Ripples in the initial condition

$$u(x, 0) = F(x) \tag{3.13}$$

lead to singularities familiar to students of catastrophe theory (Arnol'd 1984). We have, for example, with $F(x) = x^{1/\alpha}$,

$$u^\alpha + ut - x = 0 \tag{3.14}$$

as an integral of (3.11); for $\alpha = 2$ or 3 we have structurally stable catastrophes. As the singularity forms, it is to be replaced by a discontinuity before multiple values arise. The discontinuity is (in this schematic account) soon forgotten as dissipation takes its toll. The local nature of the shocks, as dictated by α, dominates the power spectrum of u:

$$P(k) = \int e^{ikx} u^2 \, dx \sim \int e^{ikx} x^{2/\alpha} \, dx \sim k^{-(1+2/\alpha)}, \tag{3.15}$$

with $\alpha = 2$ and 3 as the typical values. In the next section, I attempt to adumbrate the qualitative implications that this may have for turbulence.

4. TURBULENCE

There may be many kinds of fluid turbulence. Certainly, there are many ways to stir up a fluid. Let us focus on the example of Boussinesq convection, in which the motion first arises in a supercritical, stationary bifurcation. As the control parameter, R, increases, more and more modes become unstable and growth rates become very large.

In the nineteenth century, Lord Kelvin and others spoke of convective equilibrium, a new state that a highly unstable fluid achieves. They imagined that the instabilities produce strong motions that carry much heat. This, in modern jargon, cause a renormalization of R back to almost the marginal value; that is the new equilibrium. This picture is still influential in discussion of stellar convection theory and, in fluid dynamics, the writings of Malkus sometimes sound a similar theme.

If we examine perturbations to convective equilibrium in the spirit of the foregoing discussion, we can expect to find numerous nearly marginal modes. The competition among these modes will give us a chaotic state, just as in the mildly unstable situation. This time, however, the competition of modes is the typical situation and, in that sense, turbulence is a chaotic phenomenon.

To formulate these notions, we first assume that $\langle U \rangle$, the ensemble average of U, is not explicitly time dependent at very large R. (This assumption may be alright for convection, but is not likely to work for pipe flow; hence our choice of example.) $\langle U \rangle$ is to play the role that U_0 did in the preceding paragraphs. Though U_0 depends only on z, $\langle U \rangle$ may well depend on x, y, and z. (Elsewhere, in attempts to make this discussion quantitative, I have assumed $\langle U \rangle$ depends only on z, but that may be optimistic).

Now, we proceed at first as if $\langle U \rangle$ were known and look for solutions $U = \langle U \rangle + u$. The linear theory for small $\|u\|$ will reveal only mild instability according to Lord Kelvin's notions. But there ought to be numerous mildly unstable and mildly stable modes, and these will control the dynamics. As before,

$$U(x, t) = \langle U \rangle + \sum_i A_i(x, y, t)\, \boldsymbol{\phi}_i(x, y, z) + \text{corrections.} \qquad (4.1)$$

The asymptotic methods that I have sketched tell us the structure of the equations for A_i. The solution of such a system of equations poses the usual problem of the statistical theory of turbulence. The nuance is that the equations contain coefficients that depend upon $\langle U \rangle$. If these equations could be solved, we would simply need certain moments of the solution to compute $\langle U \rangle$ itself. This self-consistency problem does not offer a practical approach to turbulence theory, but it does allow us to organize a qualitative vision of certain turbulent flows.

I have already suggested that the qualitative validity of the idea of convective equilibrium makes chaos almost an inevitable feature of turbulence. But convective equilibrium is a mean property and a temporary breakdown of it will unleash large instabilities that will play a part in turbulent intermittency. For extended systems, we can have nonlinear packets of modes, as in §3, and these, I suggest, are the coherent structures of turbulence (Spiegel 1983).

I am not pretending that there is a theory of turbulence here. But I am suggesting that the phenomena of mildly unstable systems may be the building blocks of such a theory. Whether this is also what is meant by people who interchange the words chaos and turbulence, I cannot say. What I find a little disturbing about this, innocuous as it seems, is that it seems to be rooted in a mathematical vision.

The amplitudes A_i are not really observables. Rather, they resemble the internal variables of quantum physics. Indeed, the calculations of the normal forms of the equations for A_i involve, at bottom, finding certain group representations (Spiegel 1985). It also appears that the dynamics of the A_i involve secondary flow fields that resemble gauge fields. My picture is that much of the complication of turbulence arises in these fields, like those studied in chaos theory.

G. I. Taylor wrote that turbulence involved intense concentrations of vorticity. Others have said this since. Many have sought the origin of these concentrations in the development of singularities, especially in the Euler equations. In §3, I suggested that some near singularities occur in the complex amplitude functions of nonlinear instability theory. What I am suggesting is that we should look for the long-suspected singularities of turbulence in the evolution of the internal variables, or amplitude vectors. It may be that in a field of their shock waves we may discern the texture of turbulence.

These remarks contain snatches from various discussions and unpublished collaborations with many more people including D. W. Moore and S. Childress and, most recently, with S. Zaleski, made possible in part by a NATO travel grant. I am also indebted to the NSF for support under NSF PHY80-27321.

References

Arnéodo, A., Coullet P. H. & Spiegel, E. A. 1982 *Physics Lett.* A **92**, 369–373.

Arnol'd, V. I. 1986 *Catastrophe theory.* (Berlin: Springer.)

Bretherton, C. F. & Spiegel, E. A. 1983 *Physics Lett.* A **96**, 152–156.

Carr, J. 1981 *Application of centre manifold theory.* (*Applied Mathematical Sciences* series 35.) New York: Springer.

Coullet, P. H. & Spiegel, E. A. 1983 *SIAM Jl appl. Math.* **43**, 775–819.

Coullet, P. H. & Spiegel, E. A. 1987 In *Proc. Workshop on Energy Stability and Convection, Capri, 1986* (ed. P. Galdi & B. Straughan). (*Longman Research Notes in Maths.* (In the press.).)

Kuramoto, Y. & Tsuzuki, T. 1976 *Prog. theor. Phys.* **55**, 356–369.

Lorenz, E. N. 1963 *J. atmos. Sci.* **20**, 130–141.

Moon, H. T., Huerre, P. & Redekopp, L. G. 1982 *Phys. Rev. Lett.* **49**, 458.

Moore, D. W. & Spiegel, E. A. 1966 *Astrophys. J.* **143**, 871–887.

Newell, A. C. & Whitehead, J. A. 1969 *J. Fluid Mech.* **38**, 279–303.

Segel, L. A. 1969 *J. Fluid Mech.* **38**, 203–224.

Stuart, J. T. 1962 In *Proc 10th Ing. Cong. on Applied Mechanics, Strega, 1960*, p. 63. New York: Elsevier.

Spiegel, E. A. 1983 In *Chaos in astrophysics* (ed. J. Buchler, J. Perdang & E. A. Spiegel), pp. 91–135. Dordrecht: Reidel.

Spiegel, E. A. 1985 In *Theoretical approaches to turbulence* (ed. D. L. Dwoyer, M. Y. Hussaini & R. G. Voigt) (Applied Mathematical Science series 58). New York: Springer-Verlag.

Swift, J. & Hohenberg, P. C. 1977 *Phys. Rev.* A**15**, 319.

Arithmetical theory of Anosov diffeomorphisms

By F. Vivaldi

School of Mathematical Sciences, Queen Mary College, University of London,
Mile End Road, London E1 4NS, U.K.

Nineteenth century arithmetic is used to study periodic orbits of Anosov diffeomorphisms of the two-dimensional torus. We find that the period of the orbits, as well as their dynamical behaviour, are intimately related to the way ideals factorize in algebraic number fields.

1. Introduction

Let us consider the following area-preserving map of the two-dimensional torus:

$$r' = Ar; \quad A = \begin{bmatrix} 4 & 15 \\ 1 & 4 \end{bmatrix}; \quad \det A = 1, \quad \operatorname{tr} A = 8. \tag{1.1}$$

A is a hyperbolic system, its eigenvalues being real. It belongs to the family of Anosov diffeomorphisms, which tipify purely chaotic motion in hamiltonian systems. They are characterized by uniform hyperbolicity and a dense set of unstable periodic orbits (for a review, see Franks 1970).

Let us now consider a rational point on the torus.

$$r = (n_1/p, n_2/p), \tag{1.2}$$

where p is a prime number (except 2, 3 and 5), and n_1 and n_2 are integers between 0 and $p-1$. One finds that the orbit through r has the following properties.

(i) It is periodic and its period T depends on p alone.

(ii) T divides $p-1$ if and only if one of the following diophantine equations

$$\pm p = x^2 - 15y^2 = Q_1(x, y), \tag{1.3a}$$

$$\pm p = 3x^2 - 5y^2 = Q_2(x, y), \tag{1.3b}$$

has solutions in integers x and y. If this is not so, then T divides $p+1$.

Thus, to study periodic orbits, we are forced to make a seemingly artificial distinction between prime numbers. For instance, the prime 7 is representable by the quadratic form Q_2 ($x = 2$, $y = 1$), 11 by $-Q_1$ ($x = 2$, $y = 1$), whereas 13 has no such representation. Therefore all orbits whose coordinates are rational with denominators 7, 11 and 13, respectively, have periods dividing 6, 10 and 14, respectively.

The above situation is typical. Indeed in the theory of Anosov diffeomorphisms of the torus, one is naturally led to consider unimodular matrices acting on rational points (as long as one is interested in periodic orbits). To clarify this point, we note the following results.

[97]

1. Every Anosov diffeomorphism of the n-dimensional torus is topologically conjugate to a hyperbolic toral automorphism (Manning 1975).

So for the two-dimensional case we need only consider 2×2 matrices with integer entries (for continuity), determinant equal to 1 (for area-preservation) and real eigenvalues (for hyperbolicity).

2. Periodic orbits of hyperbolic toral automorphisms have rational coordinates, and vice versa.

This can be verified without difficulty. In particular, all rational points sharing the same denominator form a two-dimensional invariant lattice on the torus, and one is interested in classifying and constructing all its orbits. A natural approach consists in dealing with prime denominators first, and then building up the general case from prime factors.

The formulation of results about periodic orbits in terms of quadratic forms is very appealing, for it requires little specialized jargon. Moreover, using quadratic forms, one can afford to solve a class of problems in modern dynamics relying almost entirely on the techniques developed by Gauss in his *Disquisitiones arithmeticae* (Gauss 1801). The apparent simplicity of this approach is, however, misleading, especially from a computational viewpoint. We shall not pursue this matter here, but merely make the following observations.

1. The discriminant of the characteristic equation of A in (1.1): $x^2 - 8x + 1 = 0$, is equal to 60. (The reader will note that the exceptional primes 2, 3 and 5 are precisely the prime divisors of 60.)

2. There are two non-equivalent forms of discriminant 60 (actually four, counting the sign multiplicity), namely Q_1 and Q_2 defined in (1.3). (The discriminant of the quadratic form $ax^2 + bxy + cy^2$ is $b^2 - 4ac$.)

3. The primes representable by $\pm Q_1$, are not representable by $\pm Q_2$, and vice versa. The problem of characterizing primes representable by a given quadratic form is in general very difficult.

An alternative approach is to use the more abstract but computationally (and conceptually) simpler notion of ideal factorization, which was developed in the nineteenth century precisely to overcome the difficulties connected with solving diophantine equations. To prepare the reader on what to expect, we now reformulate the results concerning the system (1.1) as follows.

(i) T divides $p + 1$ if and only if p is a prime ideal (p is 'inert').

(ii) T divides $p - 1$ if and only if p is the product of two distinct prime ideals (p 'splits').

(iii) The numbers 2, 3, and 5 are the product of two identical prime ideals, i.e. they are squares (2, 3 and 5 'ramify').

All the action takes place in the quadratic field $\mathbb{Q}(\sqrt{60})$, to which the eigenvalues of A belong. The meaning of all this will become clear in the following sections.

2. QUADRATIC FIELDS

Dynamics on two-dimensional lattices calls for a generalization of the concept of integers. Ordinary integers form the monodimensional lattice $\mathbb{Z}$, which is equipped with a multiplication, i.e. it is a ring. $\mathbb{Z}$ is contained in the field $\mathbb{Q}$ of rational numbers. An example of a two-dimensional lattice with a multiplicative structure is given by the set of complex numbers with integer real and imaginary parts (gaussian integers). This set is embedded in the field of complex numbers whose real and imaginary parts are rational. Gaussian integers play no role in our theory, but they are useful for illustrative purposes.

To generalize the above example, we choose an integer D with no square divisors, and we define the quadratic field $\mathbb{Q}(\sqrt{D})$ as the aggregate

$$\mathbb{Q}(\sqrt{D}) = (z: z = a + b\sqrt{D}; \ a, b \text{ rational}). \tag{2.1}$$

One easily verifies that $\mathbb{Q}(\sqrt{D})$ is closed under all four arithmetical operations. The set $\mathbb{Z}[\omega]$ of all (quadratic) integers in $\mathbb{Q}(\sqrt{D})$ is defined as

$$\mathbb{Z}[\omega] = (z: z = m + n\omega; \ m, n \text{ integers}), \tag{2.2a}$$

where
$$\omega = \begin{cases} \sqrt{D} & \text{if } D \not\equiv 1 \pmod 4, \\ \frac{1}{2}(1 + \sqrt{D}) & \text{if } D \equiv 1 \pmod 4. \end{cases} \tag{2.2b}$$

Thus $\mathbb{Z}[\omega]$ is a two-dimensional lattice in $\mathbb{Q}(\sqrt{D})$, generated by the two basis vectors 1 and ω and integer coefficients (one writes $\mathbb{Z}[\omega] = [1, \omega]$). $\mathbb{Z}[\omega]$ is also closed under multiplication (but not division), i.e. it is a ring.

In spite of the unattractive specialization (2.2b), $\mathbb{Q}(\sqrt{D})$ and $\mathbb{Z}[\omega]$ are legitimate generalizations of $\mathbb{Q}$ and $\mathbb{Z}$, respectively. This is best seen by starting from quadratic equations with integer coefficients. It is not difficult to show that (2.1) and (2.2) are equivalent to

$$\mathbb{Q}(\sqrt{D}) = (z: az^2 + bz + c = 0, \quad \text{with} \quad b^2 - 4ac = n^2 D, D \text{ square-free}) \tag{2.3a}$$

$$\mathbb{Z}[\omega] = (z: z^2 + bz + c = 0, \quad \text{with} \quad b^2 - 4c = n^2 D, D \text{ square-free}) \tag{2.3b}$$

Replacing quadratic equations with linear ones in the definitions (2.3), one recovers $\mathbb{Q}$ and $\mathbb{Z}$, respectively.

We now need some standard definitions (here u, v, z are integers in $\mathbb{Z}[\omega]$, for some fixed ω given by (2.2b)).

Divisibility: z divides v if there exists u such that $zu = v$.

Units: u is a unit if it divides 1 (whence all integers).

Associates: z and v are associates if $z = uv$ for some unit u.

Indecomposables: z is indecomposable if $z = uv$ implies that u or v are units.

Primes: p is prime if when p divides uv then p divides either u or v.

Units play the same role as the numbers 1 and -1 in $\mathbb{Z}$. This generalization is not trivial, because in real quadratic fields there is an infinity of units (for instance, $2\,143\,295 + 221\,064\sqrt{94}$ is a unit). Note the careful distinction between indecomposables and primes, which is unnecessary in $\mathbb{Z}$ (see below).

By definition, u is a unit precisely when $c = \pm 1$ in (2.3b). It follows that eigenvalues of toral automorphisms are *units* in quadratic fields (because $c = 1$), and indeed this property is the arithmetical equivalent of area-preservation.

We are now in a position of establishing a tight relation between dynamics and arithmetic, by transforming the action of unimodular matrices on lattices into multiplication in suitable rings of quadratic integers. We start from an eigenvalue λ (the largest one, say) of a toral automorphism, and construct the ring $\mathbb{Z}[\omega]$ to which λ belongs (cf. (2.2)). We identify a rational point on the torus $(n_1/g, n_2/g)$ with the quadratic integer $z = n_1 + n_2\,\omega$, where n_1 and n_2 are to be taken modulo g. Because λ is itself a quadratic integer, so is the product $z' = \lambda z$, and therefore z' can be expressed in term of the basis of $\mathbb{Z}[\omega]$: $z' = n_1' + n_2'\omega$. Thus multiplication by λ in $\mathbb{Z}[\omega]$ induces a map on $\mathbb{Z}^2$, which carries (n_1, n_2) to (n_1', n_2'). This map is linear, and can be constructed explicitly by determining its action on the basis $[1, \omega]$ of $\mathbb{Z}[\omega]$.

$$\lambda \begin{bmatrix} 1 \\ w \end{bmatrix} = \begin{bmatrix} a & c \\ b & d \end{bmatrix} \begin{bmatrix} 1 \\ w \end{bmatrix} = A^{\mathrm{T}} \begin{bmatrix} 1 \\ w \end{bmatrix}, \tag{2.4}$$

where A^{T} is the transpose of A. Then $A(n_1, n_2) = (n_1', n_2')$, as readily verified. Moreover A has determinant 1, λ being a unit. Because λ depends just on one integer coefficient (namely b in (2.3b)), so does A, and we have obtained a one-parameter family $A = A(b)$ of hyperbolic matrices ($|b| > 2$).

The family $A(b)$ does not exhaust the whole of $\mathrm{SL}_2(\mathbb{Z})$, even if we allow for conjugacy (in $\mathrm{SL}_2(\mathbb{Z})$). The treatment of the full problem is of interest more to algebraists than dynamicists, and it will not be considered here. It is worth noting that matrices having the same eigenvalues have (essentially) the same dynamics, regardless of conjugacy considerations (see Vivaldi 1987 and references therein).

The structure of periodic orbits is closely related to that of indecomposables and primes in $\mathbb{Z}[\omega]$. Let us consider again the ring $\mathbb{Z}[\sqrt{-1}]$ of gaussian integers, and recall a classical result, the Fermat's genus theorem (1640).

An odd prime is the sum of two squares if and only if it is of the form $4n+1$.

Fermat's theorem was the first instance in which representability of primes by means of a binary quadratic form (the sum of two squares) was related to *arithmetic progressions*. We now see that it is also related to the laws of decomposition of gaussian integers, and one finds that there are three possibilities. The primes of the form $4n+1$, which are representable by the form $Q(x, y) = x^2 + y^2$, are no longer primes in $\mathbb{Z}[\sqrt{-1}]$ (they 'split'), because $p = x^2 + y^2 = (x + y\sqrt{-1})\,(x - y\sqrt{-1})$. In this case the two factors of p are gaussian primes (this is not obvious). The primes of the form $4n-1$ are not sum of two squares, and so they remain primes in $\mathbb{Z}[\sqrt{-1}]$ (they are 'inert'). Finally, $p = 2$ is an exception, in that it is the sum of two squares, but it factors as $2 = (-\sqrt{-1})\,(1 + \sqrt{-1})^2$ (2 'ramifies'). Then 2 is not prime, being an associate of the square of the prime $1 + \sqrt{-1}$ ($-\sqrt{-1}$ is a unit).

Let us now construct the 'genus' theorem for our example (1.1). The eigenvalue of A is $\lambda = 4 + \sqrt{15}$, and so the relevant ring is $\mathbb{Z}[\sqrt{15}]$ (cf. (2.2b)).

A prime is (apart from a sign) either the sum of a square minus 15 times another square, or 3 times a square minus 5 times another square, if and only if it is of the form $60n+k$, *with* k *one of* 1, 7, 11, 17, 43, 49, 53, 59. *The primes 2, 3 and 5 are exceptions.*

The simplicity and elegance of Fermat's theorem is lost. Representability of primes is still controlled by arithmetic progressions, but we can no longer relate the latter to prime factorizations in $\mathbb{Z}[\sqrt{15}]$, because more than one quadratic form is involved. If we attempt to factorize p, we find that $p = Q_1(x, y) = x^2 - 15y^2 = (x + y\sqrt{15})(x - y\sqrt{15})$ does yield factors in $\mathbb{Z}[\sqrt{15}]$, but $p = Q_2(x, y) =$

$3x^2 - 5y^2 = (x\sqrt{3} + y\sqrt{5})(x\sqrt{3} - y\sqrt{5})$ factors in another field (which is not even quadratic!). As to the exceptional primes 2, 3 and 5, none of them is an associate of a square, because they are representable by Q_2 and not Q_1.

All this depends on the failure of the fundamental theorem of arithmetic in $\mathbb{Z}[\sqrt{15}]$, where integers do not factor uniquely as the product of indecomposables. As an example, we have $6 = 2.3 = -(3 + \sqrt{15})(3 - \sqrt{15})$, and these two factorizations are irreconcilable, because all factors of 6 can be shown to be indecomposable. They are not prime though. For instance, $3 + \sqrt{15}$ divides the product of two integers, 2 and 3, but it does not divide either one. This is the rule rather than the exception in algebraic number fields.

It is the wish to recover a link between arithmetic progressions and factorization that forces us to introduce the concept of ideal. This is done in the next section.

3. Ideals

Ideals are special types of lattices, which will correspond to invariant sets on the torus. As such, they have a strong influence on the behaviour of periodic orbits. They can be multiplied and divided, and the resulting (multiplicative) arithmetic is similar to ordinary arithmetic, yet more powerful and closer to the dynamical problem. The classical reference for ideal theory is Dedekind (1871). See also Hecke (1923) or Cohn (1962).

We begin by considering ideals in $\mathbb{Z}$, which are just sets of multiples of integers. The multiples of n will be denoted by (n), a one-dimensional lattice. Conversely, any lattice in $\mathbb{Z}$ is the set of multiples of some integer. We define $(m)(n) = (mn)$. Clearly, $(-n) = (-1)(n) = (n)$, i.e. multiplication by a unit leaves an ideal invariant. The statement that m divides n means that (m), as a set, *contains* (n). This can be taken as the definition of divisibility, which is normally done using multiplication. Indeed (m) contains (n) if and only if there exists an ideal (o) such that $(m)(o) = (mo) = (n)$. Ordinary arithmetic and ideal arithmetic coincide in $\mathbb{Z}$ because the prime factorization $n = p_1^{z_1} \dots P_t^{z_t}$ corresponds to the ideal factorization of (n) into prime ideal factors: $(n) = (p_1^{z_1}) \dots (p_t^{z_t})$.

The multiples of a quadratic integer $z \in \mathbb{Z}[\omega]$ also form an ideal, called a *principal ideal*, and denoted by (z). Now (z) is a two-dimensional lattice. As before, $(z)(v) = (zv)$, and (z) divides (v) if and only if (z) contains (v). From its definition, (z) has the following properties

$$u, v \in (z) \Rightarrow u \pm v \in (z) \quad \text{(lattice property)}, \tag{3.1a}$$

$$a \in \mathbb{Z}[w], v \in (z) \Rightarrow av \in (z) \quad \text{(ideal property)}. \tag{3.1b}$$

Because (3.1a) does not imply (3.1b), not all lattices are ideals. More important, there can exist lattices in $\mathbb{Z}[\omega]$ satisfying (3.1a,b), but not being multiples of any integer. Accordingly, we shall take (3.1) as the definition of an ideal in $\mathbb{Z}[\omega]$. Ideals which are not multiples of any integer are said to be *non-principal*: they occur precisely when unique factorization fails. In any case, an ideal I, as a two-dimensional lattice, has a two-elements integral basis $I = [w_1, w_2]$, with w_1 and w_2 quadratic integers. Each element of I can be represented as $z = n_1 w_1 + n_2 w_2$, with n_1 and n_2 integers. If $J = [\omega_3, \omega_4]$ is another ideal, the product IJ is then defined as $IJ[\omega_1 \omega_3, \omega_1 \omega_4, \omega_2 \omega_3, \omega_2 \omega_4] = [u, v]$, for some $u, v \in \mathbb{Z}[\omega]$. The last equality follows because IJ, as a sublattice of $\mathbb{Z}[\omega]$, must itself admit a two-element integral basis. Also, if λ is a unit, then $(\lambda) = (1)$, since λ divides 1, and therefore $(\lambda)I = (1)I = I$ for any ideal I (multiplication by a unit merely rearranges the basis elements). This property will make ideals into invariant sets (see below).

Let us consider some examples. In $\mathbb{Z}[\sqrt{-1}]$ (gaussian integers), all ideals are principal, and ideals and integers factor in the same way. Ideals are 'square' lattices, because they must be invariant under multiplication by $\sqrt{-1}$, which is a rotation by $\frac{1}{2}\pi$ (cf. (3.1b)). According to Fermat's theorem, 3 is not the sum of two squares, and therefore it does not factor in $\mathbb{Z}[\sqrt{-1}]$. This means that $(3) = [3, 3\sqrt{-1}]$ is *contained* by no square lattice but itself and $(1) = \mathbb{Z}[\sqrt{-1}]$ (see figure 1a). The prime 5 is the sum of two squares: $5 = 1^2 + 2^2 = (2 + \sqrt{-1})(2 - \sqrt{-1})$, whence $(5) = [5, 5\sqrt{-1}]$ is contained by $(2 + \sqrt{-1}) = [2 + \sqrt{-1}, -1 + 2\sqrt{-1}]$ and $(2 - \sqrt{-1}) = [2 - \sqrt{-1}, 1 + 2\sqrt{-1}]$ (figure 1b). Finally 2 is an associate of a square in $\mathbb{Z}[\sqrt{-1}]$, which yields the ideal factorization $(2) = (1 + \sqrt{-1})^2 = [1 + \sqrt{-1}, -1 + \sqrt{-1}]^2$ (figure 1c).

In $\mathbb{Z}[\sqrt{15}]$, the factorization $-11 = Q_1(2, 1) = (2 + \sqrt{15})(2 - \sqrt{15})$ corresponds to the ideal factorization $(-11) = (11) = [11, 11\sqrt{15}] = (2 + \sqrt{15})(2 - \sqrt{15}) = [11, 2 + \sqrt{15}][11, -2 + \sqrt{15}]$, and all the ideals involved here are principal. On the other hand, the prime 7 does not factorize in $\mathbb{Z}[\sqrt{15}]$ (because it is represented by Q_2), yet (7) has two (non-principal) ideal factors: $(7) = [7, 7\sqrt{15}] = [7, 1 + \sqrt{15}][7, -1 + \sqrt{15}]$. By the same token, the exceptional primes (2), (3), and (5), are all squares of non-principal ideals: $(2) = [2, 1 + \sqrt{15}]^2$, $(3) = [3, \sqrt{15}]^2$, $(5) = [5, \sqrt{15}]^2$, whereas ideals like (13) or (19) are prime, i.e. they do not have ideal factors.

Ideals now become an essential feature of these discrete phase spaces. The ring $\mathbb{Z}[\omega]/(g)$, which represents rational points on the torus with denominator g, will in general contain non-trivial ideals, namely the ideal divisors of (g). To find them, it suffices to consider the case $g = p$, a prime. In the examples above we have identified three different ways in which (p) decomposes in a quadratic field. Thus (p) may remain prime ((p) is inert), or have two distinct ideal factors ((p) splits) or be a square ((p) ramifies). This decomposition is unique, even when factorization into indecomposable integers is not. The ideal factors of (p) are invariant under the map, because the latter corresponds to multiplication by the unit λ. This is why ideal factorization (rather than representability by quadratic forms, or factorization of integers) is the key to the understanding of the dynamics.

The relation between ideal factorization and arithmetic progressions is very deep, and also computationally useful. It allows one to characterize completely the structure of prime ideals in term of a finite number of tests, for which efficient algorithms are available. The modulus that controls the arithmetic progressions (4 and 60, respectively, in our previous examples) is called the (field) *discriminant d*. It is defined in term of the square-free kernel D as follows (compare with (2.2b))

$$d = \begin{cases} 4D & \text{if} \quad D \not\equiv 1 \pmod 4 \\ D & \text{if} \quad D \equiv 1 \pmod 4. \end{cases} \tag{3.2}$$

For a given discriminant, an arithmetical function is defined, which characterizes the splitting status of a prime. It is the so-called *Kroenecker symbol* (d/p),

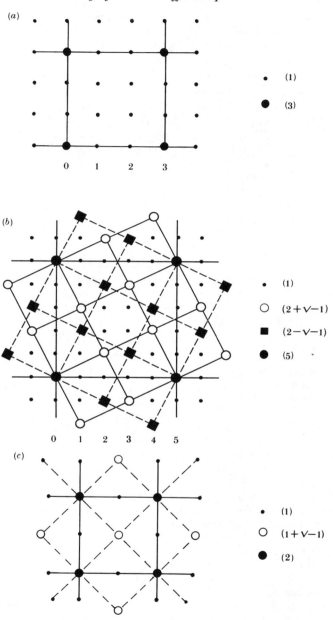

FIGURE 1. Ideal-theoretic interpretation of Fermat's genus theorem. (*a*) 3 is inert. (*b*) 5 splits: $(5) = (2 + \sqrt{-1})(2 - \sqrt{-1})$. (*c*) 2 ramifies: $(2) = (1 + \sqrt{-1})^2$.

which assumes the values -1, 1, and 0, respectively, depending whether p is inert, splits or ramifies in $\mathbb{Q}(\sqrt{d})$ (see Cohn 1962). We can now summarize prime factorization in quadratic fields as follows:

$$
\begin{array}{llll}
(p) = (p) & \text{iff} & (d/p) = -1 & p \text{ is inert,} \\
(p) = P_1 P_2 & \text{iff} & (d/p) = +1 & p \text{ splits,} \\
(p) = P_1^2 & \text{iff} & (d/p) = 0 & p \text{ ramifies.}
\end{array}
\tag{3.3}
$$

The Kroenecker symbol can then be extended to composite 'denominators', by the laws $(d/mn) = (d/m)(d/n)$, and $(d/-n) = (d/n)$. The important fact is that (d/n), as a function of n, is *periodic with period d*, which brings about arithmetic progressions modulo the discriminant. To prove this, and to compute (d/n) efficiently, one must make full use of such concepts as quadratic residues and reciprocity (Gauss 1801).

Inert and split primes form infinite families (Dirichlet 1837), so that the dynamical behaviour that characterizes them (see §4) will also occur infinitely often. On the other hand, $(d/p) = 0$ only if p divides d (p prime), whence, for a given discriminant, the ramified primes are finite in number.

4. Results on periodic orbits

In this section we summarize the most important results concerning periodic orbits. We shall not provide proofs (which will be found in Percival & Vivaldi 1987), but rather emphasize the role played by various parts of the theory. The meaning of some keywords of arithmetic will be assumed known throughout this section.

Prime moduli

Our original purpose was to determine the period of orbits with a given prime denominator p. If λ is the eigenvalue of the map, and z the initial condition, this is done by determining the smallest integer T such that

$$
\lambda^T z \equiv z (\text{mod } (p)),
\tag{4.1}
$$

where this congruence means that we are working with $\mathbb{Z}[\omega]/(p)$. The integer z may or may not belong to an ideal divisor of (p) (if any). In the former case, the whole orbit through z will remain on such ideal (which is invariant), and we shall speak of an *ideal orbit*. In the latter, the orbit is said to be *free*. In correspondence to the three possible factorizations of (p), we have three different orbit behaviours, as detailed below (the fixed point $z = 0$ is ignored).

1. If $(d/p) = -1$, (p) has no factors. All orbits are free and have the same period T, which is a divisor of $p+1$. If $T = (p+1)/m$, then there are $m(p-1)$ free orbits.

2. If $(d/p) = 1$, $(p) = P_1 P_2$ has two ideal factors. All orbits have the same period T, which divides $p-1$. If $T = (p-1)/m$, then there are $m(p-1)$ free orbits, and $2m$ ideal orbits.

3. If $(d/p) = 0$, then $(p) = P_1^2$. The periods of orbits is computed as follows. Let $\lambda = \frac{1}{2}(a+b\sqrt{D})$ (with a and b both even if $D \not\equiv 1 (\text{mod } 4)$). We have two cases (when $p = 2$ the first case always holds).

(i) If $a \equiv 2(\bmod p)$, there are $p-1$ ideal orbits of period 1, and $p-1$ free orbits of period p (or $p(p-1)$ of period 1). Each free orbit belongs to the same residue class modulo P_1.

(ii) If $a \equiv -2(\bmod p)$, there are $\frac{1}{2}(p-1)$ ideal orbits of period 2, and $\frac{1}{2}(p-1)$ free orbits of period $2p$ (or $\frac{1}{2}p(p-1)$ of period 2). Each free orbit belongs to two residue classes mod P_1.

Note that in general the exact period cannot be established *a priori*, and some computation is required. Because the average number of divisors of a large integer n grows like $\ln(n)$, the average number of possible periods that needs be tested is $\ln\left(\frac{1}{2}(p+1)\right)$.

The structure of the ideal factors of p does not affect just the period, but also the dynamical behaviour of orbits. Orbits of inert primes have no distinctive features. In the split case there are ideal orbits, which are confined to invariant sublattices. Each such lattice contains $p-1$ points (excluding the origin), which are uniformly distributed on the torus. It follows that ideal orbits having maximal period (i.e. $p-1$) visit all points on the ideal, and therefore they are also uniformly distributed. Ramified primes support orbits of peculiar regularity. The ideal orbits contain fixed points, while free orbits are strongly correlated. Among these orbits the so-called accelerator modes (Chirikov 1979) are to be found.

Initial conditions

Points on ideal orbits are easy to find, from a knowledge of the basis of the ideal. To find a representative point for free orbits, we must instead return to quadratic forms. The form involved is the *principal form* of discriminant d

$$Q(x,y) = \begin{cases} x^2 - (\frac{1}{4}d)y^2, & d \text{ even}, \\ x^2 + xy - [\frac{1}{4}(d-1)]y^2, & d \text{ odd}. \end{cases}$$

If we write $z = x + yw$, then $Q(z) = Q(x,y)$, when taken modulo p, is invariant under the map, because λ is a unit. $Q(z)$, which is also called the *norm* of z, then becomes a property of the whole orbit. For simplicity, let the period be maximal (i.e. $T = p+1$, $p-1$, and $2p$, depending on cases). Then it can be shown that the orbits are uniquely identified by the value of their norm, and to find a point on an orbit one must find a quadratic integer z of appropriate norm. We are again dealing with diophantine equations, except that we are now working modulo p. One makes use of the following result (cf. Hecke 1923, theorems 138–139).

If $(d/p) \neq 0$ the congruence $Q(z) \equiv a(\bmod p)$ is solvable in z for any a, whereas if $(d/p) = 0$ it is solvable if and only if a is a quadratic residue modulo p.

Now if $Q(z)$ is a residue (non-residue) modulo p, so is $Q(kz) = k^2 Q(z)$ (k integer), because it differs from it by a square factor. Moreover the numbers $Q(kz)$, $k = 1, \ldots, \frac{1}{2}(p-1)$ are all distinct modulo p. This means that the points kz (k as above) belong to different orbits. Thus initial conditions will be found on two monodimensional lattices, generated by any two quadratic integers z_1 and z_2 whose norm is a quadratic residue and (if $(d/p) \neq 0$) a non-residue modulo p, respectively. The most obvious choice for z_1, is 1, whereas z_2 will in general be found by educated trial and error.

Composite moduli

The case of a composite modulus (g) is solved in two stages. First one deals with the case in which g is a primary integer, $g = p^n$ (p prime), and then one combines results from different primary factors. Let $g = p^n$. Then the period of the orbits increases very regularly (indeed linearly) with n, beyond a certain power n_0. We have the following results.

1. Let T be the period of free orbits to the prime modulus (p), and n_0 the smallest integer n for which $\lambda^T \not\equiv 1 (\mathrm{mod}\ (p^n))$. Then, for odd primes p, the period T_n of all free orbits modulo (p^n) is the same: $T_n = Tp^{n-n_0+1}$. If $p = 2$ the period is one of the numbers $T, Tp, Tp^2, \ldots, Tp^{n-n_0+1}$.

2. Let T_n and n_0 be as above. Then orbits on any ideal factor of (p^n) which is not divisible by (p) have the same period $T_n^1 = T_n/a$ where $a = a(n)$ is either 1 or p. If p is an odd prime, then $a(n) = a(n_0 + 1)$, for all $n > n_0$.

In the general case of a composite modulus $(g) = (p_1^{\alpha_1}) \ldots (p_t^{\alpha_t})$, one must first determine the periods T_k for all corresponding primary factors $(p_k^{\alpha_k})$. Then the period of the orbits for the modulus (g) (and the same initial condition) is just the least common multiple of the T_ks, from the Chinese remainder theorem. One can also show that all free orbits have the same period, as do ideal orbits belonging to conjugate ideal factors of (g).

Orbits of given period

The set of orbits of a prescribed period n is characterized by the smallest denominator M common to all of their points. $M = M(n)$ is called the *maximal modulus* for the period n. Not all points with denominator M need have period (dividing) n. Those which do turn out to form an ideal divisor $O = O(n)$ of (M) (see below). In order to construct all orbits of a given period, one needs to compute M, O, and then locate orbits in O using the techniques described above.

The following three propositions characterize M and O.

1. Let the norm of $\lambda^n - 1$ be equal to $\pm m^2 r$, with r square-free. Then the maximal modulus is $M = mr$. Moreover, if g divides m, then all points with denominator g have period dividing n, and vice versa.

2. All orbits whose period divides n form a principal ideal $O = O(n)$, which divides the corresponding maximal modulus $M(n)$. We have $O = (\sqrt{D})$ for even n, whereas when n is odd, $O = ((\lambda - 1)/m_1)$, where m_1 is the largest rational integer dividing $\lambda - 1$. Then O is an ambiguous ideal, i.e. it is equal to its conjugate and has no rational divisors.

3. Let $\lambda > 0$. Then $O = (1)$ if and only if the period is odd and $\lambda = \eta^2$, where η is a unit of norm -1.

When $O = (1)$ all points with denominator M have period (dividing) n, a welcome simplification. The occurrence of this phenomenon is related to the existence of units norm -1, a classical problem which is still not completely solved (see Cohn 1978, p. 105).

There is a recursive formula for the computation of the maximal modulus, which causes M to increase exponentially with the period n. For instance, in the example (1.1) we have $M(20) = 3543553200$, which is the denominator one must consider to construct all orbits of period 20.

The significance of this cannot be overestimated. Orbits of Anosov diffeomorphisms are exponentially unstable, yet arithmetic is free from instabilities and numerical errors. In integer dynamics, the unavoidable difficulties inherent to computing with chaotic motions manifest themselves in the appearance of exponentially large integers.

In closing we would like to remark that Anosov diffeomorphisms of the n-dimensional torus can be treated similarly, using suitable algebraic fields of degree n.

I thank Professor I. Percival for many stimulating discussions, and S.E.R.C. for supporting this research.

REFERENCES

Chirikov, B. V. 1979 *Physics Rep.* **52**, 263.
Cohn, H. 1962 *A second course in number theory.* New York: John Wiley & Sons (Reprinted as *Advanced number theory.* New York: Dover (1980).)
Cohn, H. 1978 *A classical invitation to algebraic numbers and class fields.* New York: Springer Verlag.
Dedekind, R. 1871 Über die Theorie des ganzen algebraischen Zahlen (Suppl. II Dirichelet, P. G. L. *Vorlesungen über Zahlentheorie.* Braunschweig (1894).)
Dirichlet, P. G. L. 1837 *Abh dt. Akad. Wiss. Berl. (Werke* **1**, 315.)
Dirichlet, P. G. L. 1894 *Vorlesungen über Zahlentheorie.* Braunschweig. (Repr. New York: Chelsea Pub. Co. (1968)).
Franks, J. 1970 Anosov diffeomorphisms. In *Proc. Symp. Pure Math.*, vol. 14, p. 61. Providence, Rhode Island: *American Mathematical Society.*
Gauss, C. F. 1801 *Disquisitiones arithmeticae.* Leipzig. (English transl. Yale, New Haven and London (1966).)
Manning, A. 1975 Classification of Anosov diffeomorphisms on tori. *Lecture Notes in Mathematics*, no. 468. New York: Springer-Verlag.
Hecke, E. 1923 *Vorlesung Über die Theorie der Algebraischen Zahlen.* Leipzig: Akademische Verlagsgesellschaft. (English trans. *Lectures on the theory of algebraic numbers* (New York: Springer-Verlag (1981).)
Percival, I. C. & Vivaldi, F. 1987 Arithmetical properties of strongly chaotic motions. *Physica* D **25**, 105.
Vivaldi, F. 1987 The Arithmetic of Chaos. In *Fractals and Chaos.* Bristol: Institute of Physics. (In the press.)

Chaotic behaviour in the Solar System

By J. Wisdom

Department of Earth, Atmospheric, and Planetary Sciences, Massachusetts Institute of Technology, Cambridge, Massachusetts 02139, U.S.A.

There are several physical situations in the Solar System where chaotic behaviour plays an important role. Saturn's satellite Hyperion is currently tumbling chaotically. Many of the other irregularly shaped satellites in the Solar System had chaotic rotations in the past. There are also examples of chaotic orbital evolution. Meteorites are most probably transported to Earth from the asteroid belt by way of a chaotic zone. Chaotic behaviour also seems to be an essential ingredient in the explanation of certain non-uniformities in the distribution of asteroids. The long-term motion of Pluto is suspiciously complicated, but objective criteria have not yet indicated that the motion is chaotic.

1. Introduction

The Solar System is generally perceived as evolving with clockwork regularity. Indeed, it was a search for the principles that underlie the perceived regularities in the motions of the planets that culminated in Newton's formulation of the laws of mechanics and universal gravitation 300 years ago. Recently it has been widely recognized that dynamical systems possess irregular as well as regular solutions. Irregular solutions of deterministic equations of motion are termed 'chaotic'. The Solar System is just another dynamical system; the study of this preeminent dynamical system is not untouched by the discoveries in nonlinear dynamics. Solar System dynamics encompasses the orbital and rotational dynamics of the planets and their natural satellites, the coupling between them, and the slow evolution of the orbits and spins due to tidal friction. It is primarily the dynamics of resonances and resonances are almost always associated with chaotic zones. Chaotic behaviour must be considered a possibility in almost any dynamical situation in the Solar System. In this paper a number of physical applications of modern dynamics to the Solar System will be reviewed. Applications to rotational dynamics will be considered first, followed by applications to orbital dynamics.

2. Tumbling of Hyperion

The chaotic tumbling of Hyperion, one of Saturn's more distant satellites with an orbit period of 21 days, offers one of the most dramatic physical examples of chaotic behaviour (Wisdom *et al.* 1984). The rotation rate and spin-axis orientation are predicted to undergo significant changes in only a few orbit periods. The chaotic tumbling of Hyperion is primarily a consequence of Hyperion's highly aspherical shape, which was determined from *Voyager 2* images to have radii of

190 km × 145 km × 114 km ± 15 km (Smith *et al.* 1982), and to a lesser extent a consequence of the large eccentricity of Hyperion's orbit ($e \approx 0.1$). Weak tidal friction acting over the age of the Solar System is responsible for bringing Hyperion to this chaotic state.

An out-of-round satellite in a non-uniform gravitational field is subject to a torque. The torque arises because the attractive force on the side of the satellite nearest the planet is stronger than the attractive force on the far side of the satellite. For an out-of-round body the torques arising from these forces do not balance, and give rise to a net torque, a 'gravity gradient torque'. Hyperion is subject to especially large torques because of its highly aspherical shape. In addition, these torques have a strong time dependence because of the large eccentricity of Hyperion's orbit.

Earth's Moon very nearly always points the same face toward Earth; the equality of the rotation period and the orbit period of the Moon is a natural consequence of the action of tidal friction. Tidal friction tends to bring the spin axis into coincidence with the axis of largest moment of inertia, and over longer times brings the spin axis perpendicular to the orbit plane as the rotation rate is slowed until the rotation period equals the orbital period (see Goldreich & Peale 1966; Peale 1977). All satellites in the Solar System which are sufficiently close to their host planet for the tidal torques to have been strong enough to significantly affect the rotation rate over the age of the Solar System are observed to be in this state where the spin period is locked to the orbit period. The timescale for the spin of Hyperion to be slowed by tidal friction to synchronous rotation is on the order of the age of the Solar System. Thus, the magnitude of Hyperion's rate of rotation is near that which Hyperion would need to always point one face toward Saturn. Hyperion's rotation would not be chaotic if it were not tidally evolved. At the same time, if the timescale for tidal despinning were much shorter than the age of the Solar System, Hyperion might have already found its way into a stable commensurate rotation state.

The chaotic rotation of Hyperion is best illustrated in a simplified model. In this model the orbit of Hyperion is taken to be a fixed ellipse; the timescale for chaotic variations in the spin rate is much shorter than the timescale for significant variations in Hyperion's orbit, which are, in any case, not large. Furthermore, the spin axis is taken to be perpendicular to the orbit plane and aligned with the axis of largest moment of inertia; this is the usual outcome of tidal evolution. In this simplified problem the equation of motion for the orientation is quite simple:

$$C\frac{\mathrm{d}^2\theta}{\mathrm{d}t^2} = -n^2(B-A)\frac{3}{2}\left(\frac{a}{r}\right)^3 \sin 2(\theta-f).$$

The orientation of the satellite is specified by a single angle, θ, which is taken to be the angle between the axis of smallest principal moment of inertia (the longest axis of a triaxial ellipsoid) and the inertially fixed line of periapse (the line joining the planet and the point in the orbit closest to the planet). The angular position of the satellite in its orbit is also measured from the periapse of the orbit. This angle is the true anomaly, denoted here by the symbol f. The principal moments of inertia are $A < B < C$; C is the moment of inertia about the spin axis. The mean angular motion of the satellite in its orbit is n, the instantaneous distance from the planet to the satellite is r, and the semimajor axis of the orbit is a. The equation of motion equates the rate of change of the angular momentum, or equivalently, the product of the moment of inertia about the spin axis and the acceleration of the orientation, to the external torque. The inverse cube dependence of the torque on the distance from the planet reflects the origin of the torque as a gravity gradient. There would be no net torque if the body were axisymmetric about the spin axis; the asymmetry of the body enters the equation of motion through the difference of the principal moments of inertia in the plane of the orbit, $B-A$. Instantaneously, the torque always tends to try to align the long axis of the satellite with the line between the satellite and the planet; the angle between the long axis and the planet–satellite line is $\theta-f$. This equation keeps only the lowest moments of the mass distribution in the orientation dependent part of the potential energy. The contributions that are ignored are of one higher order in the small ratio of the radius of the satellite to the orbital radius. In this approximation all bodies have a symmetry under which a rotation by 180° gives a dynamically equivalent configuration. The factor of two multiplying the difference of angles reflects this symmetry.

This equation of motion has only a single degree of freedom, the orientation angle θ, but depends explicitly on the time through the distance to the planet, r, and the non-uniform keplerian motion of the true anomaly, f. It is worth emphasizing that it is the non-zero eccentricity of the orbit that spoils the integrability of this simplified problem. If the eccentricity is set to zero then the planet to satellite distance remains equal to the semimajor axis, and the true anomaly becomes simply the mean motion times the time. The equation of motion for the angle $\theta' = \theta - nt$ is

$$C\frac{\mathrm{d}^2\theta'}{\mathrm{d}t^2} = -\frac{3n^2(B-A)}{2}\sin 2\theta'.$$

Except for the factor of two, which could easily be removed by a further change of variables, this is the equation of motion for a pendulum, which of course can be explicitly integrated. An important feature is that the problem now has an integral

$$E = \tfrac{1}{2}C\left(\frac{\mathrm{d}\theta'}{\mathrm{d}t}\right)^2 - \frac{3n^2(B-A)}{4}\cos 2\theta'.$$

Hamiltonian systems with more than one degree of freedom almost always exhibit a divided phase space: for some initial conditions the trajectory is chaotic, and for others the trajectory is regular (Hénon & Heiles 1964). The explicit time-dependence cannot be eliminated from the equations of motion for the simplified spin-orbit problem when the eccentricity is non-zero. Thus the spin-orbit problem may be expected to display the generic mixed phase space. The structure of the phase space is most easily understood by computing surfaces of section. For the simplified spin-orbit problem surfaces of section are generated by looking at the rotation state stroboscopically, once per orbit. The equation of motion is numeri-cally integrated, and every time the satellite goes through periapse the rate of change of the orientation, $d\theta/dt$, is plotted against the orientation, θ. The surface of section for Hyperion is shown in figure 1. A number of different trajectories have been used to illustrate the principal types of motion that are possible. Recall that if the motion is quasiperiodic the points will fill a one-dimensional curve; if the points seem to fill an area the motion is chaotic. All of the scattered points in

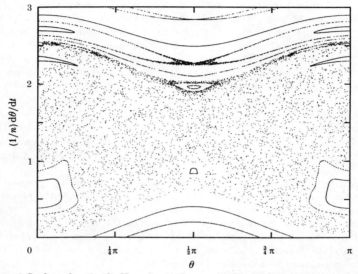

FIGURE 1. Surface of section for Hyperion (with $\alpha = \sqrt{[3(B-A)/C]} = 0.89$ and $e = 0.1$). The rate of change of the orientation is plotted against the orientation at every periapse passage. The spin axis is fixed perpendicular to the orbit plane.

the centre of the section belong to the same trajectory. The two trajectories that generate an X in the upper centre part of the section are also chaotic. The other trajectories appear, without close examination, to be quasiperiodic, and certainly identify the main regions of quasiperiodic motion. The islands in the chaotic sea correspond to various states where the rotation period is commensurate with the orbit period. The island in the lower part of the section near $\theta = 0$ is the synchronous island, where Hyperion would on the average always point one face toward Saturn (i.e. never make a complete relative rotation). The island in the upper part of the large chaotic zone is the 2-state, where Hyperion would on the average rotate twice every orbit period. A number of other islands are shown. The curves in the bottom of the section near $\theta = \frac{1}{2}\pi$ represent a non-commensurate quasiperiodic rotation. If the range of the ordinate were greater it would be seen that they stretch all the way across the figure, as do other non-commensurate quasiperiodic curves near the top of the section. Only the portion of the section between 0 and π is shown because the addition of π to θ gives a dynamically equivalent state. Note, however, that the synchronous states on the section at $\theta = 0$ and $\theta = \pi$ differ in that they present opposite faces to Saturn.

This simplified model was motivated by the standard picture of the tidal evolution of rotations in which the spin axis is driven to the orbit normal as the spin is slowed to synchronous rotation. Elements of the standard picture must now be reexamined. In particular, it is necessary to reexamine the stability of the spin axis orientation perpendicular to the orbit plane. Without giving the details of the methods used, it turns out that the chaotic zone is attitude unstable. This means that if Hyperion were placed in the chaotic zone with the slightest deviation of its spin axis from the orbit normal this deviation would grow exponentially. The timescale is just a few orbit periods. This is also true of the synchronous state; that state in which all other tidally evolved satellites in the Solar System are found is attitude unstable for Hyperion! The attitude stability of the other commensurate islands is mixed, some are stable while others are unstable. The equations that govern the three-dimensional tumbling motion are Euler's equations with the full three-dimensional gravity gradient torque. These equations have three degrees of freedom, through, say, the three Euler angles, plus the explicit time-dependence from the non-uniform keplerian motion in an orbit with non-zero eccentricity. It is no longer possible to plot a surface of section for a problem with so many degrees of freedom. However, another property of chaotic trajectories is that neighbouring trajectories separate exponentially from one another. The rates of exponential separation are quantified by the Lyapunov characteristic exponents. The three-dimensional tumbling state which is entered as the spin axis falls away from the orbit normal is a fully chaotic state. There are no hidden integrals of the motion; the chaotic tumbling motion has three positive Lyapunov exponents.

When the evolution due to tidal friction is included the problem is no longer strictly hamiltonian. However, there is a tremendous disparity between the dynamical timescale and the timescale over which the tides are important. The tidal evolution is consequently viewed as a slow evolution through the phase space of the hamiltonian system. Most likely Hyperion at one time had a rotation period much shorter than its orbital period and began its evolution high above the top of the section in figure 1. Over the age of the Solar System its spin gradually slowed, while the obliquity damped nearly to zero. As it damped to zero the assumptions made in computing figure 1 came closer to being realized. By the time Hyperion reached the large chaotic zone its spin axis was nearly normal to the orbit plane. Once the large chaotic zone was entered, however, the work of the tides over aeons was undone in a matter of days. Because the large chaotic zone is attitude unstable, Hyperion quickly began to tumble through all orientations. Ultimately, Hyperion may be captured by one of the small attitude stable islands. It can never be captured by the synchronous island because the synchronous island is attitude unstable.

Observations of Hyperion are not yet adequate to fully confirm the chaotic tumbling, though they are all consistent with it. The most convincing evidence for chaotic tumbling comes from the *Voyager* pictures themselves, which show that the long axis of Hyperion is out of the orbit plane and the spin axis is near the plane. This is consistent with chaotic tumbling, but inconsistent with other known regular rotation states. Further observations of Hyperion will be needed to unambiguously determine whether its rotation is chaotic. If the observations are complete enough it might be possible to invert the light curve for the initial conditions and moments. Numerical simulations indicate that the moments may be determined with accuracy which increases exponentially with the time interval over which the observations are made.

3. IRREGULARLY SHAPED SATELLITES

Are there other examples of chaotic tumbling in the Solar System? Hyperion appears to be alone in its chaotic dance, the result of a unique combination of factors which are nowhere else realized in the Solar System. It turns out though that many other satellites tumbled chaotically in the past. In fact, *all* irregularly shaped satellites in the Solar System must tumble chaotically just at the point where the spin is about to be captured into synchronous rotation (Wisdom 1987a).

Almost all resonances are surrounded by chaotic zones, though in some cases these chaotic zones may be very narrow. The commensurate spin-orbit states are examples of resonances. Resonances appear as islands on a surface of section. Two moderately narrow chaotic zones were illustrated in the upper part of figure 1. There exist approximate methods of estimating the size of these chaotic zones (see Chirikov 1979). The width of the chaotic zone surrounding the synchronous island may be specified in terms of the magnitude of the chaotic variations of the integral E of the zero eccentricity problem:

$$\frac{\Delta E}{E} \approx \frac{14\pi e}{\alpha^3}\, \mathrm{e}^{-\pi/2\alpha},$$

where α, the asphericity parameter, is $\sqrt{[3(B-A)/C]}$. Note that this estimate has the correct limit for zero orbital eccentricity, where the simplified problem is integrable. Although the width of the chaotic zone depends exponentially on the asphericity parameter, it only depends linearly on the orbital eccentricity. Thus satellites with large deviations from spherical symmetry, but small eccentricities may still have significant chaotic zones.

Phobos, a satellite of Mars, is almost as out-of-round as Hyperion, but its orbital eccentricity is only 0.015. A surface of section for Phobos is shown in figure 2. The chaotic zone is a significant feature on the section. Even for Deimos, the other satellite of Mars, where the orbital eccentricity is considered to be anomalously small ($e \approx 0.0005$), the chaotic zone is not microscopic (see figure 3.) Surfaces of section for several other irregularly shaped satellites with α near unity confirm the existence of significant chaotic zones surrounding the synchronous island.

Stability analysis shows that the chaotic zones of these irregular satellites is in every case attitude unstable, just as it is for Hyperion. A slight displacement of the spin axis from the orbit normal grows exponentially, leading to chaotic tumbling. The surprising result is the strength of this attitude instability. In every case the timescale for the exponential growth of obliquity is only a few orbit

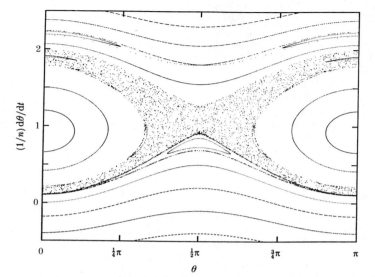

FIGURE 2. Surface of section for Phobos (with $\alpha = 0.83$ and $e = 0.015$). The chaotic zone is a significant feature on the section.

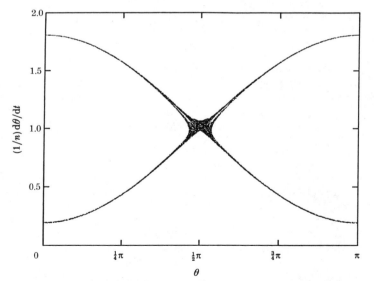

FIGURE 3. Chaotic separatrix for Deimos (with $\alpha = 0.81$ and $e = 0.0005$). The chaotic zone is sizable considering the very low orbital eccentricity.

periods. This is true even for Deimos, with its low orbital eccentricity, and narrow chaotic zone. The orbital eccentricity does not play a crucial role in this attitude instability. It turns out that even for zero eccentricity the synchronous separatrix is attitude unstable, and leads to chaotic three-dimensional tumbling, even though the simplified problem with the spin axis perpendicular to the orbit plane is integrable.

It is not possible to tidally evolve into the synchronous state without passing through a region that is attitude unstable. The resulting tumbling motion is always chaotic. All synchronously rotating satellites with significantly irregular shapes must have spent a period of time tumbling chaotically. The length of time spent in this state must be comparable with, and probably somewhat greater than, the despinning timescale; it is not yet possible to make a rigorous estimate. Thus Deimos probably spent of the order of 100 Ma tumbling chaotically, and Phobos spent on the order of 10 Ma in this tumbling state.

Enhanced dissipation of energy during the chaotic tumbling phase may help explain the anomalously low eccentricity of Deimos, and certainly must be taken into account in future studies of the orbital histories of the irregularly shaped satellites. This new episode in the adolescence of the irregularly shaped satellites is, however, fascinating in itself. The World, and newtonian mechanics in particular, works in a surprising way.

4. 3/1 Kirkwood gap

The distribution of the semimajor axes of the asteroids is not uniform; it shows several gaps as well as several enhancements. The origin of these gaps has been the object of a great deal of speculation. One major clue to the cause of these non-uniformities, which was noted at the time of their discovery by D. Kirkwood, is that they occur near mean-motion commensurabilities with Jupiter. That is, a small integer times the mean motion of an asteroid in a gap will nearly equal the product of another small integer times the mean motion of Jupiter. However, the mere association of a gap with a resonance does not in itself explain the formation of the gap. Nature herself provides the counterexamples: there are gaps at some resonances and enhancements at others. The basic difficulty in understanding the formation of the gaps was that the motion near complex resonances was not well understood analytically, and numerical simulations could not alleviate the problem because of the great amount of computer time required. Integrations over 10 ka did not uncover a mechanism for the formation of gaps. Integrations over significantly longer times were prohibitively expensive and did not seem warranted.

Longer integrations were made possible by the introduction of a new method for following the trajectories of asteroids (Wisdom 1982). Following the ideas of Chirikov (1979), an algebraic mapping of the phase onto itself was derived which approximates motion near the 3/1 commensurability. The map is an approximation to the stroboscopic section that would be obtained by looking at the coordinates of the asteroid once each Jupiter period (which is about 12 years); evolution of an asteroid is followed by successively iterating this map. The derivation of the map relies on the averaging principle, as have most other studies of the long-term evolution of asteroid orbits. The terms with highest frequency, the orbital frequency, are first removed by averaging, leaving the resonant terms and the secular terms. New high-frequency terms are added in such a way that delta functions are formed. The new equations can be integrated across the delta functions and between them, giving a map of the phase space onto itself. The map is significantly faster than more conventional methods; it is more than a thousand times faster than a full integration and several hundred times faster than the methods that rely on numerical averaging to increase the basic step size. This great increase in computation speed made integrations over much longer intervals possible.

Integration over longer times was justified. Fig. 4 shows the orbital eccentricity as a function of time for a chaotic trajectory near the 3/1 commensurability

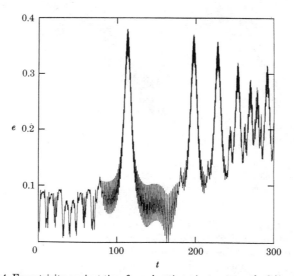

FIGURE 4. Eccentricity against time for a chaotic trajectory near the 3/1 commensurability. Time is measured in millennia. A short (10 ka) integration would give a very poor idea of the nature of this trajectory.

computed with the planar elliptic map. Although excursions in eccentricity of this magnitude were previously known (Scholl & Froeschlé 1974), the possibility that an orbit could spend a hundred thousand years or longer at low eccentricity and then 'suddenly' take large excursions was quite unexpected. Subsequent numerical integrations of the full, unaveraged, differential equations have verified that the behaviour is not an artifact of the method (Wisdom 1983; Murray & Fox 1984).

It is only when the trajectory is computed over millions of years that one begins to feel as though the true nature of the motion is represented. Figure 5 shows the typical behaviour of the eccentricity of a chaotic trajectory near the 3/1 resonance in the planar elliptic problem. There are bursts of irregular high-eccentricity behaviour interspersed with intervals of irregular low-eccentricity behaviour, with

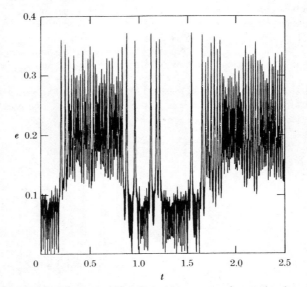

FIGURE 5. Eccentricity of a typical chaotic trajectory over a longer time interval. The time is now measured in millions of years. Bursts of high-eccentricity behaviour are interspersed with intervals of irregular low-eccentricity behaviour, broken by occasional spikes.

an occasional eccentricity spike. Figure 6 shows a very interesting, though relatively rare behaviour. The eccentricity jumps shown in figure 6 all reach the same eccentricity, but seem to occur at irregular intervals. A most surprising result is that if the plot is expanded near two different jumps and then superimposed the eccentricity jumps are practically identical. Both of these trajectories were computed with the map. For discussions of the growth of numerical error see Wisdom (1983, 1987 b).

The unexpected behaviour of the eccentricity can be understood by putting the trajectory in context on a surface of section (Wisdom 1985 a). At first sight this is not possible because the planar elliptic problem has two and a half degrees of freedom (through, say, the x and y coordinates and the explicit time dependence resulting from the keplerian motion of Jupiter in its elliptical orbit), but the problem may be reduced to two degrees of freedom by averaging over the orbital period. A surface of section corresponding to figure 5 is shown in figure 7. Here the variables $x = e \cos{(\varpi - \varpi_J)}$ and $y = e \sin{(\varpi - \varpi_J)}$ are plotted each time a particular combination of the mean longitudes goes through zero. The distance from the

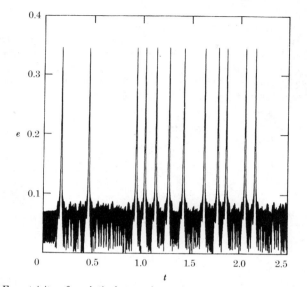

FIGURE 6. Eccentricity of a relatively rare, but quite interesting chaotic trajectory. The time is measured in millions of years.

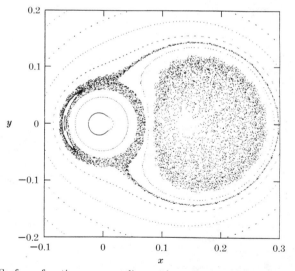

FIGURE 7. Surface of section corresponding to the trajectory of figure 5. The coordinates are $x = e \cos(\varpi - \varpi_J)$, and $y = e \sin(\varpi - \varpi_J)$. The orbital eccentricity is the radius from the origin. The trajectory is free to explore a rather large chaotic zone, but sometimes spends a period of time near the islands close to the origin.

origin on this section is the orbital eccentricity, e, and the ϖs are the longitudes of perihelia for the asteroid and Jupiter. During the intervals of low-eccentricity behaviour the trajectory stays in that part of the chaotic zone near the origin, encircling one of the islands; during the high eccentricity intervals the trajectory is moving in the extended chaotic zone to the right of the figure. The origin of the peculiar behaviour of the eccentricity shown in figure 6 is apparent on the surface of section shown in figure 8. The chaotic zone that surrounds the origin has a very

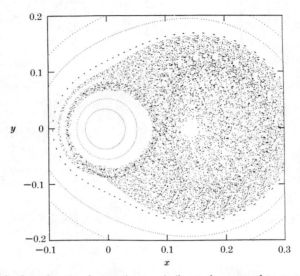

FIGURE 8. Surface of section for a trajectory similar to the one used to generate figure 7. The narrow branch of the chaotic zone explains the irregularly appearing, but nearly identical jumps in eccentricity.

narrow branch that extends to high eccentricity. The similarity of the different jumps is explained by the narrowness of the chaotic zone. The fact that the jumps occur at irregular intervals simply reflects the irregular nature of the motion in the chaotic zone. Thus the peculiar behaviour of the eccentricity of chaotic trajectories near the 3/1 commensurability can be understood as a simple manifestation of chaotic behaviour in a problem with two degrees of freedom.

The long-period motion can also be understood semianalytically (Wisdom 1985 b). Over much of the phase space the resonance timescale and the secular timescale are well separated. An analytic average over the resonance timescale gives a long-period hamiltonian with one degree of freedom. This approximation, for instance, recovers the large jumps in eccentricity. It also gives a rather interesting picture of the evolution in the chaotic zone, where the chaotic trajectories are for the most part predictable, but occasionally enter a region where the motion is essentially four dimensional.

Now that the nature of trajectories near the 3/1 commensurability is better understood, can the formation of the 3/1 Kirkwood gap be explained? The large eccentricity increases are important for the formation of the gap because at the location of the gap eccentricities above 0.3 are Mars-crossing. It turns out that all of the chaotic trajectories cross the orbit of Mars. Orbits that previously appeared to be limited to low eccentricity are now understood to have large excursions in eccentricity on longer timescales. The quasiperiodic resonance librators also generally have sufficient variation in eccentricity to cross the orbit of Mars. Thus asteroids with both resonant quasiperiodic trajectories and chaotic trajectories near the 3/1 commensurability can be removed by close encounters or collisions with Mars. Comparison of the outer boundary of the chaotic zone with the actual distribution of asteroids shows remarkably good agreement (figure 9). This figure gives strong evidence that chaotic behaviour has indeed played a role in the formation of this gap.

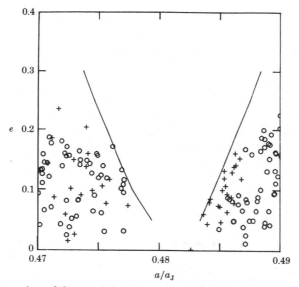

FIGURE 9. Comparison of the actual distribution of asteroids with the outer boundaries of the chaotic zone. There is both a chaotic region and quasiperiodic region in the gap, but trajectories of both types are planet-crossing.

5. TRANSPORT OF METEORITES

It is rather surprising that the origin of the meteorites, those stones which contain so many clues concerning the formation of the Solar System, is still not definitively known. It is widely believed that meteorites originate in the asteroid belt, yet until recently a dynamical mechanism for transporting them to Earth which was consistent with the meteorite date eluded discovery. Monte-Carlo simulations of Wetherill (1968) ruled out the previously suggested sources, and seemed to indicate that there was another source of high-eccentricity orbits.

When the orbit of Jupiter is assumed to be fixed, and the motion of the asteroid is limited to Jupiter's orbit plane, chaotic trajectories near the 3/1 commensurability are limited to eccentricities below about 0.4. As the integration is made more realistic, though, by allowing three-dimensional motion and including the variations of Jupiter's orbit which result from the perturbations of the other planets, the variations in eccentricity become more extreme. Chaotic trajectories which begin at normal asteroidal eccentricities ($e \approx 0.15$) reach eccentricities above 0.6, which is large enough for the orbit to cross the orbit of Earth. Figure 10 shows an example of such behaviour. Besides giving a stronger mechanism for clearing the 3/1 Kirkwood gap, these chaotic trajectories provide a new dynamical mechanism for bringing debris from asteroidal collisions near the 3/1 resonance directly to Earth (Wisdom 1985b). Wetherill (1985) has shown that this new source is consistent with the meteorite data, and that the larger fragments from asteroid collisions partly account for the observed population of Earth crossing asteroids. This discovery of a dynamical route from the asteroid belt to Earth is an important scientific application of chaotic behaviour.

6. 2/1 KIRKWOOD GAP AND THE HILDA ASTEROIDS

The fact that there is a gap in the distribution of asteroid semimajor axes near the 2/1 commensurability and an enhancement in the distribution near the 3/2 commensurability needs an explanation. Unfortunately, the dynamics of the 2/1 and 3/2 resonances are considerably more complicated that the dynamics of the 3/1 resonance. These resonances are not well represented by low-order truncations of the disturbing potential. This makes analytic investigations difficult if not impossible.

The only alternative appears to be direct numerical integrations. Numerical integration of problems in celestial mechanics is particularly time consuming because of the great range of timescales involved. The orbital dynamics of the Solar System only begins to be interesting when studied over timescales of millions of years. The single trajectory in figure 10 used the equivalent of about 200 VAX hours. The Digital Orrery (Applegate *et al.* 1985) is a special purpose computer specifically designed to study problems in celestial mechanics. The design and construction of the Orrery were led by Gerald J. Sussman, from the Artificial Intelligence Laboratory and the Department of Electrical Engineering at MIT. The construction of the Orrery is an extremely important advance for planetary dynamics, which evidently had a need for a dedicated supercomputer. The Orrery runs at about 60 times the speed of a VAX for celestial mechanics problems, or about a third the speed of a Cray.

Chaotic behaviour near the 2/1 commensurability was first discovered by Giffen (1973), though short integrations by Froeschlé & Scholl (1976, 1981) indicated that chaotic behaviour was not very widespread or catastrophic. A new survey of the structure near the 2/1 and 3/2 resonances is currently underway

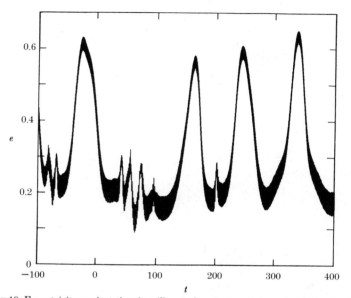

FIGURE 10. Eccentricity against time in millennia for a test particle perturbed by the four jovian
planets. At times the eccentricity is typical of asteroids, at other times it is large enough
for the orbit to cross the orbit of Earth. This integration shows meteoritic material may
be directly transported to Earth from the asteroid belt by way of the 3/1 chaotic zone.

with the Digital Orrery. The preliminary results of this survey are quite interesting. Figure 11 shows the chaotic zone near the 2/1 commensurability. In this
figure a cross indicates that a trajectory of the planar elliptic problem with this
initial eccentricity and semimajor axis is chaotic. Though the exploration is not
yet complete, the outline of the chaotic region is probably well represented. There
appears to be a sizable chaotic zone near the 2/1 commensurability. On the other
hand, the corresponding plot for the region near the 3/2 commensurability shows
that the resonance region is basically devoid of chaotic behaviour. There is thus
a *qualitative* difference in the structure of the phase space near the 2/1 and 3/2
commensurabilities which corresponds to the qualitative difference in the observed distribution of asteroids. There is a large chaotic zone at the 2/1 resonance
and there is also a Kirkwood gap at that resonance. The Hildas are located near
the 3/2 resonance and this region is devoid of chaotic behaviour. However, detailed
comparison with the actual distribution of numbered asteroids near the 2/1
resonance, figure 12, does not show perfect agreement. This discrepancy is most
likely a result of the use of the planar-elliptic approximation in the survey.
The integration of a test-particle with initial conditions in the discrepant region
perturbed by the four jovian planets showed it to be chaotic.

The association of chaotic behaviour with a gap in the distribution of asteroids
does not by itself explain the formation of the gap. For the 3/1 resonance an

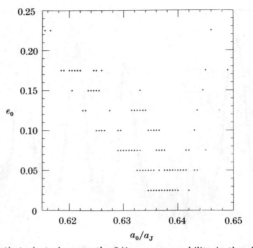

FIGURE 11. Chaotic trajectories near the 2/1 commensurability in the planar elliptic approximation. A cross marks those initial conditions (eccentricity e, and semimajor axis a, referred to Jupiter's semimajor axis) which lead to chaotic behaviour. The survey is not complete, but the basic extent of the chaotic zone is apparent. There is a significant chaotic zone.

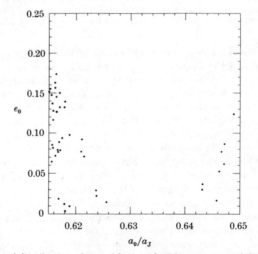

FIGURE 12. Actual distribution of asteroids near the 2/1 commensurability, each evolved to the same longitudes used in the survey. There is a good qualitative agreement between the gap and the region of chaotic behaviour shown in figure 14. The boundaries on the high semimajor axis side are in excellent agreement, whereas there seems to be a discrepancy on the low semimajor axis side.

essential ingredient was the fact that chaotic trajectories, as well as resonant quasiperiodic trajectories, crossed the orbits of Mars and Earth. The sweeping action of these planets can clear the 3/1 gap. It seems *a priori* unlikely that a similar mechanism can account for the formation of those gaps that are significantly more distant from Mars and Earth. Froeschlé & Scholl (1981) attempted to answer this question. They integrated Giffen's chaotic trajectory in the planar elliptic approximation for 100 ka, but found that it seemed to be limited to eccentricities below 0.15, which is no larger than the eccentricity of a typical asteroid. As before, the planar elliptic problem is not an adequate representation of the problem. The integration of test particles perturbed by the jovian planets shows more dramatic increases in eccentricity. For these trajectories the eccentricity and the inclination show the remarkable correlation exemplified in figure 13. Initially the inclination is low and the eccentricity seems to be limited to values below about 0.25. Over the span of the integration though the trajectory seems to trace out a pathway to high eccentricity which temporarily takes it through inclinations as high as 0.44 rad. Thus the three-dimensional nature of the motion is crucial. At the peak in eccentricity this trajectory is marginally

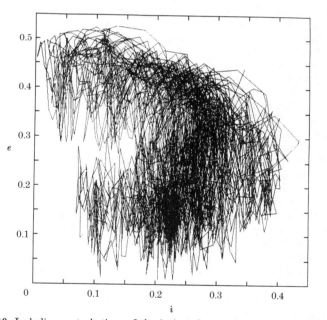

FIGURE 13. Including perturbations of the jovian planets, chaotic trajectories near the 2/1 commensurability show this remarkable correlation between eccentricity and inclination. There is a path in the phase space that takes the trajectory from low eccentricity to high eccentricity which requires that it temporarily take moderate inclination. The three-dimensional nature of the motion is crucial.

Mars-crossing. This does provide a mechanism for the removal of the asteroids on these chaotic trajectories, but I am not convinced that this is the final answer.

A few words must be said to clear up some misunderstandings. It is often said that the Kirkwood gaps are a 'simple consequence of the breakdown of K.A.M. tori near resonances'. This is not so. Resonances occur in the circular restricted problem, but the microscopic chaotic zones associated with them could never account for the creation of gaps. Such a global statement about the stability of resonances can also not account for the contrasting stability of the Hilda asteroids and the Kirkwood gaps. Only a detailed examination of successively more realistic representations of the dynamics begins to account for the distribution of asteroids. Chaotic behaviour near resonances also has nothing to do with the formation of gaps in Saturn's rings. The gaps have an entirely different origin in the collective response to resonant perturbation of a large number of particles which collide frequently, on the order of 20 times per orbit period. The qualitative character of the long-period evolution of individual trajectories is irrelevant.

7. Outer planets and Pluto

The determination of the stability of the Solar System is one of the oldest problems in dynamical astronomy. While Arnol'd's proof of the stability of a large measure of solar systems with sufficiently small planetary masses, eccentricities, and inclinations marks tremendous progress towards a rigorous answer to this question (Arnol'd 1961), the stability of the actual Solar System remains unknown. Certainly, the great age of the Solar System demands a high level of stability, but weak instabilities may still be present. Experience with the motion of asteroids has demonstrated that weak instabilities may sometimes even lead to sudden, dramatic changes in orbits. The stability of the Solar System should thus not be taken for granted.

The first application of the Digital Orrery was to the long-term evolution of the outer planets. For many years the million-year integration of Cohen et al. (1973) held the title of the longest integration of the Solar System. With the Orrery, the interval of integration has been extended to 210 Ma (Applegate et al. 1986).

The integrations showed that the best analytic approximations of the motion of the outer planets were in serious need of improvement. Bretagnon (1974) lists over 200 corrections to the Lagrange solutions. It turns out that there are contributions to the motion of the jovian planets which are larger than all but seven of those corrections. Higher-order terms were more important than the terms taken into account by Bretagnon. More recent work, particularly that of Laskar (1986, and unpublished work) is in better agreement, and provides independent confirmation of the results of the numerical integration.

The motion of Pluto is extraordinarily complicated. Pluto's orbit is unique among the planets. It is both eccentric ($e \approx 0.25$) and inclined ($i \approx 16°$). The orbits of Pluto and Neptune cross one another, a condition which is only permitted by the libration of a resonant argument associated with the 3/2 mean motion commensurability. This resonance assures that Pluto is at aphelion when Pluto and Neptune are in conjunction and thus prevents close encounters. The next level of complexity is that the argument of perihelion of Pluto librates about $\frac{1}{2}\pi$ with a period near 3.8 million years (Williams & Benson 1971). The Orrery integrations confirmed this libration, but found that the picture was not yet complete; there are significant contributions to the motion of Pluto with much longer periods. The amplitude of libration of the argument of perihelion has a strong modulation with a period of 34 million years. In fact, the frequency of the second largest contribution to the eccentricity of Pluto corresponds to a period of 137 Ma. This long period results from a near commensurability between the frequency of circulation of the longitude of the ascending node of Pluto and one of the fundamental frequencies in the motion of the jovian planets. Figure 14 shows the inclination of Pluto over 214 Ma. There seems to be an even longer period present (or perhaps even a secular drift)! The motion of Pluto is suspiciously complicated. However, the computation of the Lyapunov characteristic exponent for Pluto does not yet show any objective evidence for chaotic behaviour. Much longer integrations seem to be required to determine the true nature of Pluto's motion.

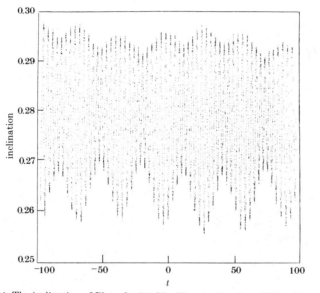

FIGURE 14. The inclination of Pluto for 214 Ma. Time is given in millions of years. Besides the 34 Ma modulation of the 3.80 Ma oscillation, there is evidence of much longer period variations (or perhaps even a secular drift!).

CONCLUSIONS

Several physical examples of chaotic behaviour in the Solar System have been presented. Hyperion tumbles irregularly as a consequence of its out-of-round shape, large orbital eccentricity, and tidally evolved rotation. Hyperion is currently the only example of this chaotic tumbling in the Solar System. However, *all* of the tidally evolved, irregularly shaped satellites in the Solar System tumbled chaotically in the past, just at the point of entry into the synchronous rotation state.

Physical examples of chaotic orbital behaviour have also been presented. The distribution of asteroids seems to be, in several instances, a reflection of the character of the trajectories in the underlying phase space. This is clearly the case for the 3/1 Kirkwood gap. There is a sizable chaotic zone, and the phase space boundary of the distribution of asteroids corresponds quite well with the outer boundary of the chaotic zone. In this case, the fact that both chaotic and quasi-periodic trajectories cross planetary orbits explains the removal of any asteroids originally in the gap.

Allowing three-dimensional motion and taking into account the perturbations of the outer planets, trajectories in the 3/1 chaotic zone reach Earth-crossing eccentricities. These trajectories seem to provide the long-sought dynamical route for the transport of meteoritic material from the asteroid belt to Earth. Studies by Wetherill have shown this source to be consistent with the ordinary chondrite data.

The 2/1 Kirkwood gap and the Hilda group have long presented a paradox to classical dynamical astronomy. A new survey with the Digital Orrery indicates that the qualitative difference in the distribution of asteroids at these two resonances is reflected in a qualitative difference in the underlying dynamics. Discrepancies in the detailed comparison probably result from the use of the planar elliptic approximation in the survey. When perturbations of the jovian planets are taken into account and three-dimensional motion is allowed, chaotic trajectories at the 2/1 resonance reach very large eccentricities at low inclinations, by way of a path that temporarily takes them to high inclinations. The three-dimensional aspect of the problem is essential. The eccentricities become large enough that the chaotic trajectories cross the orbit of Mars, but there may yet be other mechanisms for clearing the distant gaps.

The stability of the Solar System itself has been examined through a 210 Ma integration of the outer planets. The motion of the jovian planets themselves seems to be regular, though perhaps a bit more complicated than might have been expected. On the other hand, the motion of Pluto is extraordinarily complicated. Besides the well-understood mean-motion resonance which prevents the close approach of Pluto and Neptune even though their orbits cross, Pluto participates in at least two other resonances. First, it has been known for some time that the argument of perihelion librates about $\frac{1}{2}\pi$. Then, the frequency of the circulation of Pluto's ascending node is nearly commensurate with one of the fundamental frequencies in the motion of the jovian planets. This near commensurability gives rise to strong variations in the eccentricity with a period of 137 Ma. There is also evidence of much longer periods in the inclination, which appears to be secularly declining over the 210 Ma integration. Although the abundance of resonances raises suspicions about the stability of Pluto, there is not yet any objective evidence that the motion of Pluto is chaotic.

References

Applegate, J. F., Douglas, M. R., Gursel, Y., Hunter, P., Seitz, C. & Sussman, G. J. 1985 A digital orrery. *IEEE Trans. Comput.* (September issue.)

Applegate, J. F., Douglas, M. R., Gursel, Y., Sussman, G. J. & Wisdom, J. 1986 The outer solar system for 200 million years. *Astron. J.* **92**, 176–194.

Arnol'd, V. I. 1961 Small denominators and the problem of stability in classical and celestial mechanics. In *Report to the 4th All-Union Mathematical Congress Leningrad*, pp. 85–191.

Bretagnon, P. 1974 Termes a longues periodes dans le systeme solaire. *Astron. Astrophys.* **30**, 141–154.

Chirikov, B. V. 1979 A universal instability of many dimensional oscillator systems. *Phys. Rep.* **52**, 263–379.

Cohen, C. J., Hubbard, E. C. & Oesterwinter, C. 1973 Elements of the outer planets for one million years. *Astr. Pap., Wash.* **22**, 1.

Froeschlé, C. & Scholl, H. 1976 On the dynamical topology of the Kirkwood gaps. *Astron. Astrophys* **48**, 389–393.

Froeschlé, C. & Scholl, H. 1981 The stochasticity of peculiar orbits in the 2/1 Kirkwood gap. *Astron. Astrophys* **93**, 62–66.

Giffen, R. 1973 A study of commensurable motion in the solar system. *Astron. Astrophys* **23**, 387–403.

Goldreich, P. & Peale, S. J. 1966 Spin-orbit coupling in the solar system. *Astron. J.* **71**, 425–438.

Hénon, M. & Heiles, C. 1964 The applicability of the third integral of motion: some numerical experiments. *Astron. J.* **69**, 73–79.

Laskar, J. 1986 Secular terms of classical planetary theories using the results of general theory. *Astron. Astrophys.* **157**, 59–70.

Murray, C. D. & Fox, K. 1984 Structure of the 3:1 Jovian resonance: a comparison of numerical methods. *Icarus* **59**, 221–233.

Peale, S. J. 1977 Rotation histories of the natural satellites. In *Planetary satellites* (ed. J. Burns), pp. 87–112. Tucson: University of Arizona Press.

Scholl, H. & Froeschlé, C. 1974 Asteroidal motion at the 3/1 commensurability. *Astron. Astrophys.* **33**, 455–458.

Smith, B. *et al.* 1982 A new look at the Saturn system: the Voyager 2 images. *Science, Wash.* **215**, 504–537.

Wetherill, G. W. 1968 Stone meteorites: time of fall and origin. *Science, Wash.* **159**, 79–82.

Wetherill, G. W. 1985 Asteroidal source of ordinary chondrites. *Meteoritics* **18**, 1–22.

Williams, J. G. & Benson, G. S. 1971 Resonances in the Neptune-Pluto system. *Astron. J.* **76**, 167–177.

Wisdom, J. 1982 The origin of the Kirkwood gaps: A mapping for asteroidal motion near the 3/1 commensurability. *Astron. J.* **87**, 577–593.

Wisdom, J. 1983 Chaotic behaviour and the origin of the 3/1 Kirkwood gap. *Icarus* **56**, 51–74.

Wisdom, J. 1985*a* A perturbative treatment of motion near the 3/1 commensurability. *Icarus* **63**, 272–289.

Wisdom, J. 1985*b* Meteorites may follow a chaotic route to Earth. *Nature, Lond.* **315**, 731–733.

Wisdom, J. 1986 Canonical solution of the two critical argument problem. *Celest. Mech.* **38**, 175–180.

Wisdom, J. 1987*a* Rotational dynamics of irregularly shaped satellites. *Astron. J.* (Submitted.)

Wisdom, J. 1987*b* Chaotic dynamics in the Solar System. *Icarus.* (In the press.)

Wisdom, J., Peale, S. J. & Mignard, F. 1984 The chaotic rotation of Hyperion. *Icarus* **58**, 137–152.

Chaos in hamiltonian systems

By I. C. Percival, F.R.S.

School of Mathematical Sciences, Queen Mary College,
University of London, London E1 4NS, U.K.

Modern developments in hamiltonian dynamics are described, showing
the change of view that has occurred in the last few decades. The
properties of mixed systems, which exhibit both regular and chaotic
motion are contrasted with those of the integrable systems, for which the
motion is entirely regular, and of Anosov systems, for which it is almost
everywhere chaotic. The K.A.M. theorem and problems of convergence
are discussed.

Introduction

Traditionally hamiltonian systems with a finite number of degrees of freedom
have been divided into those with few degrees of freedom, which were supposed to
exhibit some kind of regular ordered motions, and those with large numbers of
degrees of freedom for which the methods of statistical mechanics should be
used.

The past few decades have seen a complete change of view, which affects almost
all the practical applications. The motion of a conservative hamiltonian system is
usually neither completely regular nor properly described by the methods of
statistical mechanics. A typical system is mixed: it exhibits regular or chaotic
motion for different initial conditions, and the transition between the two types
of motion, as the initial conditions are varied, is complicated, subtle and beautiful.

The nature of the regular motion in a system of N degrees of freedom is the
same as that of the traditional integrable systems; when bounded it is quasiperiodic, with a discrete set of frequencies ν_j, together with their integer linear
combinations:

$$\sum_{j=1}^{N} n_j \nu_j \quad (n_j \text{ not all zero}).\tag{1}$$

The regular motion for a given initial condition is confined to an N-dimensional
region in the $2N$-dimensional phase space. These regions are N-dimensional tori,
each of which can be parameterized by a set of N angle variables.

By contrast, the nature of the chaotic motion is still not fully understood. It is
unstable in a strong exponential sense. For a conservative system it usually
cannot be confined to any smooth region of dimension less than $2N-1$, the confinement required by energy conservation.

But it does not normally occupy the whole of an energy shell as required by the
ergodic principle of traditional statistical mechanics. Far away from regular regions, the chaotic motion resembles a diffusion process, but close to them it does
not. Chaotic motion is common for conservative systems of two degrees of
freedom.

Chaotic motion appears in dynamical systems when local exponential divergence of trajectories is accompanied by global confinement in the phase space. The divergence produces a local stretching in the phase space, but, because of the confinement, this stretching cannot continue without folding. Repeated folding and refolding produces very complicated behaviour that is described as chaotic.

For hamiltonian systems the stretching in one direction is exactly compensated by shrinking in another direction, so that the area in the phase space is preserved: even for $N > 1$ area preservation is fundamental, and the preservation of volume, as in Liouville's theorem, follows from it. In dissipative systems there is no such compensation. There can be chaotic motion for both, but it differs in detail. Chaotic motion may appear to be random at all times, or it may appear to be regular over long periods of time.

Poincaré was interested in the application of hamiltonian dynamics to the motion of the bodies of the Solar System, and was the first to see something of this new view of dynamics (Poincaré 1899). He recognized in particular that hamiltonian systems can show qualitatively different behaviour for arbitrarily small changes in a parameter. On the convergence of a particular series for a system of two degrees of freedom he wrote (Poincaré 1896):

> I have shown that the irrational ratios of the periods can be separated into two categories: those for which the series converge, and those for which the series diverge, and that in any interval, however small it may be, there are values from the first category, and values from the second.

Integer combinations of frequencies of the form (1) can be found with arbitrarily small magnitudes for any such interval, and these give rise to the small divisors that plague the perturbation theory of nonlinear hamiltonian dynamics. Physically, there are resonances arbitrarily close to any orbit.

But Poincaré was unable to convey his magnificent vision to his contemporaries.

The recent advances have been more closely related to other applications. For hamiltonian dynamics, these include the motion of particles in accelerators and storage rings, of stars in galaxies, of atoms in molecules and the shape of magnetic field lines in plasma containment devices.

Problems in the foundations of statistical mechanics have also played an important role, particularly concerning the nature of purely chaotic systems.

According to Chirikov (this symposium), the first published example of a system with both regular and chaotic motion appears to be that by Goward (1953) and Hine (1953). Symon & Sessler (1956) carried out computer experiments on particle orbits in accelerators, and observed 'scattered points about a bucket', which would be known today as chaotic motion about a regular island. They observed that in a particular instance two islands or resonances overlapped and that there was global chaos.

Chirikov and his collaborators (Chirikov 1979), showed that resonance overlap could be used to estimate the onset of global chaos, a general and remarkably simple method of determining this transition in systems of many degrees of freedom.

AREA-PRESERVING MAPS

Hamiltonian systems such as the Solar System, particles confined by electromagnetic fields and atoms in molecules are complicated mixed systems, with many properties that are not fully understood, so to improve this understanding we study the simplest types of system which have these properties.

These are the area-preserving maps, which are obtained by subjecting the free motion of a system in one dimension to an infinite sequence of similar impulses at unit time intervals.

The most studied area-preserving map is the standard map, introduced by Taylor (1969; see also Froeschlé 1970) and Chirikov (1969, 1971), and the subject of much detailed investigation by Chirikov and by Greene (1979). The standard map is the discrete time analogue of the vertical pendulum, in which the continuous gravitational force is replaced by a sequence of impulses at equal time intervals.

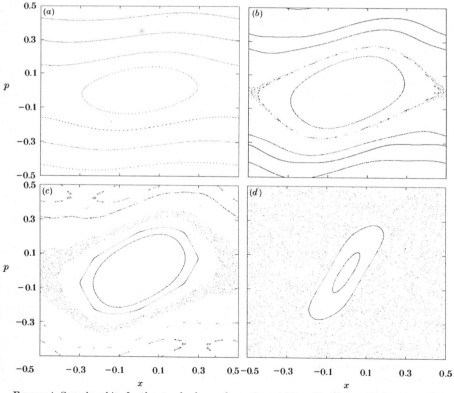

FIGURE 1. Sample orbits for the standard map for various values of k: (a) $k = 0.3$; (b) $k = 0.7$; (c) $k = 0.971\,635$; (d) $k = 3.0$.

The state of the standard map at integer time t is represented by a coordinate x, where $2\pi x$ is an angle of rotation and a conjugate (angular) momentum p. Orbits satisfy the discrete-time Hamilton equations

$$\left.\begin{aligned} p_{t+1} &= p_t - (k/2\pi) \sin (2\pi x), \\ x_{t+1} &= x_t + p_{t+1}. \end{aligned}\right\} \tag{2}$$

The parameter k represents the strength of the impulses. For $k = 0$ the motion is uniform rotation, and typical orbits for higher values of k are represented in figure 1. The structure has more detail than can be represented in any figure, with regular and chaotic motion mixed together on arbitrarily small scales.

An adequate theory of mixed systems must explain all this detailed structure: it is not yet available. The standard map illustrates the properties of a wide class of systems, obtained by perturbation of integrable systems, with the perturbation dependent on a single real parameter k. A rough picture of these systems is illustrated in figure 2.

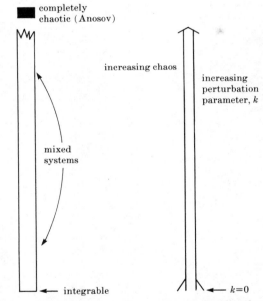

FIGURE 2. Diagram representing behaviour of some maps as a function of a parameter k, where $k = 0$ for an integrable system, and the motion becomes more chaotic when k increases. The diagram is intended as a guide to the study of non-integrable maps and should not be taken too literally.

Integrable and completely chaotic (Anosov) systems are relatively well understood. Traditional analytical dynamics illustrates the former, whereas the latter are the subject of much of modern mathematical ergodic theory, and of the paper by Vivaldi (this symposium). In between we have some knowledge of critical behaviour through the recent developments in renormalization theory, as discussed by Rand (this symposium).

These three situations are used as bases from which to explore the mixed systems.

K.A.M. AND PERTURBATION EXPANSIONS

These are used to explore the neighbourhood of the integrable systems.

The theorem of Kolmogorov, Arnol'd and Moser (K.A.M.) proves the existence of significant regular motion in some mixed systems.

K.A.M. THEOREM. *For sufficiently smooth hamiltonians, sufficiently small perturbations of nonlinear integrable systems preserve positive measure of regular motion.*

The nonlinearity requires that the derivatives of the frequencies with respect to the actions should not be too close to zero. Because the frequencies are the derivatives of the hamiltonian H with respect to the actions I_j, the condition of nonlinearity requires that

$$\det \left(\partial^2 H / \partial I_j \, \partial I_l \right) \tag{3}$$

should be bounded away from zero.

Arnol'd's smoothness condition required that the hamiltonian should be analytic in a strip in the angle and action variables, whereas Moser's original condition required it to have 333 derivatives. The smoothness condition is required to ensure that the higher Fourier components of the motion are kept under control.

Hénon (1966) carried out the computations necessary to determine the size of the perturbation in relevant units, and showed that for a system of two degrees of freedom, Arnol'd's form of the theorem required it to be less than 10^{-333} and Moser's less than 10^{-48}. The latter is less than the gravitational perturbation of a pendulum in London by the motion of a bacterium in Australia.

Numerical estimates suggest that invariant tori exist for perturbations of the order of unity for many systems, and more recent forms of the K.A.M. theorem have considerably improved on the original for specific systems (Herman 1983, 1986; Celletti *et al.* 1987). Despite its numerical inadequacies, the K.A.M. theorem has had a profound influence on the whole of hamiltonian dynamics, including the applications.

Both the K.A.M. theorem and many numerical methods of estimating invariant tori use generalized Newton–Raphson or other superconvergent iterative algorithms. For practical problems with many degrees of freedom, these methods are relatively difficult to use, so power series expansions in a perturbation parameter k are still commonly used. Furthermore, the convergence of these expansions tells us about analyticity of the motion in complex k, so, despite the K.A.M. theorem, they are still worth studying.

Let a hamiltonian system, such as the standard map, be dependent on the parameter k and integrable when $k = 0$. If an invariant torus of a given frequency exists, then it can be expressed parametrically in terms of the angle variable(s):

$$X_k(\theta) = X_k(\theta_1, \ldots, \theta_N), \tag{4}$$

where X represents a point in the phase space. Thus for the standard map:

$$X_k(\theta) = (x_k(\theta), \, p_k(\theta)). \tag{5}$$

For this case the torus is a curve, parameterized by one angle variable θ.

Do the perturbation power series in k converge for such tori? It depends on the type of perturbation theory. The identity of a torus as k is varied has to be specified. It can be identified by the action

$$I = (I_1, \, ..., \, I_N) \tag{6}$$

or by the frequency

$$\nu = (\nu_1, \, ..., \, \nu_N) \tag{7}$$

but not by both.

Consider fixed action perturbation theory. For a nonlinear system, there are normally resonances for all frequencies ν which satisfy

$$\sum_{j=1}^{N} n_j \nu_j = 0 \tag{8}$$

for any set of integers $\{n_j\}$ that are not all zero. Such frequencies give rise to the small divisors which produce the divergences of perturbation expansions, recognized by Poincaré. For fixed I, there are such frequencies for arbitrarily small k, as illustrated crudely in figure $3a$, so the power series have zero radius of convergence in k: fixed action perturbation is at best asymptotically convergent.

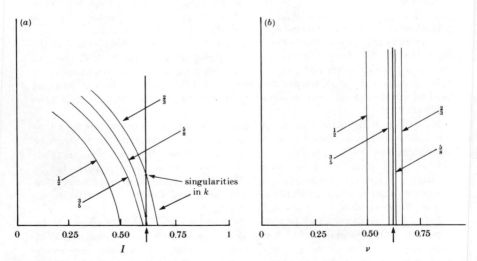

FIGURE 3. Singularities as a function of perturbation parameter: (a) for fixed action; (b) for fixed frequency. The fractions are rational approximants to the golden ratio.

Unmodified canonical perturbation theory, which is very conveniently carried out using Lie series methods, is unfortunately of this type.

On the other hand, fixed frequency perturbation theory can have a finite radius of convergence. The corresponding illustration is figure 3b. For frequencies that do not satisfy (8), the expansions do not have singularities in the frequency for any value of k. The small divisors can still give rise to large Fourier components, which impede the convergence of the series, but need not prevent it (Moser 1974).

In this, Poincaré's judgement appears to have failed him. In *New methods of celestial mechanics*, vol. 2 (Poincaré 1967) we find in ch. 13, §2, p. 91, the following statement concerning the fixed frequency expansion for a system of two degrees of freedom.

> For simplification, let us assume that two degrees of freedom exist. Could then the series, for example, converge when x_1^0 and x_2^0 (the unperturbed coordinates) are selected in such a manner that the ratio n_1/n_2 (of frequencies) becomes incommensurable and that its square, conversely, becomes commensurable ... ?
>
> The reasoning in this chapter does not permit a definite statement that this will never occur. All we can say now is that it is highly improbable.

The brackets in this quotation are not Poincaré's, and are inserted for clarity.

The most robust invariant torus is the one that survives for the largest value of k. Resonances tend to destroy tori, so one would expect the most robust torus to be furthest from the resonances, in some sense. For area-preserving maps the mapping frequency is unity, so the condition (8) is equivalent to the rationality of the frequency ν.

The golden ratio is given by

$$\tfrac{1}{2}(\sqrt{5}-1) = 1/(1+1/(1+...))...$$
$$= 0.618... \qquad (9)$$

and there is a sense in which this is the most irrational number between 0 or 1, so it is not surprising that Greene (1979) found that one of the two most robust invariant curves for the standard map is that with frequency equal to the golden ratio.

STANDARD MAP

Greene used numerical estimates of the stability of periodic orbits with frequency close to the golden ratio to determine the critical value k_c of the perturbation parameter k, such that an invariant curve exists for $k \leqslant k_c$, but not for $k > k_c$. He found that

$$k_c = 0.971\,635. \qquad (10)$$

Greene & Percival (1981) made numerical estimates of the radius of convergence of the fixed frequency perturbation expansion of the same invariant curve of the standard map, and found it to be equal to (10) to six figures. Thus according to the numerical evidence, for this case the perturbation expansion converges right up to the critical value of k, and the radius of convergence signals the breakup of the invariant curve.

The semistandard map is like the standard map, but with the sine function replaced by a single complex exponential. Series calculations are much simpler with the semistandard map, because the Fourier expansion in the angle variable θ and the perturbation expansion in the perturbation parameter k combine to form a single Taylor expansion in the variable

$$u = k\, e^{i\theta} \qquad (11)$$

(Greene & Percival 1981; Percival 1982). As a result, for any map with this property, analyticity in the perturbation parameter k is equivalent to analyticity in the angle variable θ, which has been proved to exist for a strip around the real axis in Arnol'd's (1961, 1963) form of the K.A.M. theorem.

Figure 4 illustrates the Fourier coefficients or perturbation coefficients b_n for the golden invariant curve of the semistandard map. The peaks are at the Fibonacci values of n, where n is a denominator of a rational continued fraction approximant of the golden ratio. They are caused by small divisors, and illustrate the difficulty of determining the convergence of series expansions in a field like celestial mechanics, where the difficulties of carrying out the expansion are very much greater.

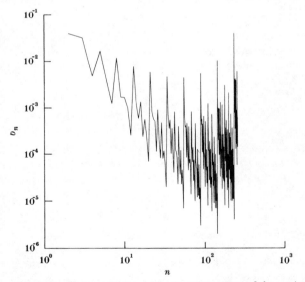

FIGURE 4. Fourier coefficients for the golden invariant curve of the semistandard map (from Greene & Percival 1981).

The similarity of the figure to that of Libchaber (this symposium), showing the behaviour of a fluid with a golden frequency ratio, is striking. It illustrates the general rule that the early terms in a fixed frequency expansion are determined mainly by the dynamics, and the later terms by number theory.

Figure 5 illustrates the radius of convergence of the fixed frequency expansion as a function of frequency. It is zero at the rationals, and for a set of irrationals (of zero measure), but nevertheless the envelope of the function is very well defined (Percival 1982). The resonances correspond to the Arnol'd tongues of the mapping of the circle to itself. It is believed (see Rand, this symposium) that the critical invariant circles of arbitrary irrational frequency exhibit universal behaviour.

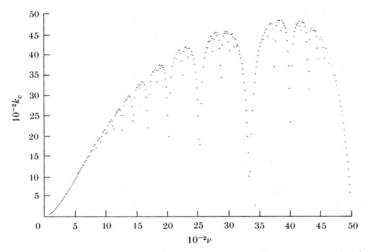

FIGURE 5. Computed radius of convergence of the fixed frequency expansion for the semistandard map, as a function of frequency (from Percival 1982).

K.A.M. is not the only rigorous theory for invariant tori. It is much easier in practice to prove the non-existence of tori than to prove their existence, because for the latter one has to take account explicitly of all resonances, whereas for the former one does not.

By a considerable generalization of some results of Birkhoff (1920, 1932), theorems on non-existence of tori have been proved by Aubry *et al.* (1982), Aubry (1983), Hermann (1985) and Mather (1982a, 1984). Based on these theorems, and their extension, some non-rigorous tests of feasibility by Newman & Percival (1983), were followed by MacKay & Percival's (1985) computer proof of the non-existence of the golden invariant circle of the standard map for $k > \frac{63}{64} = 0.984375$, which is not far from the computer estimates of the critical $k = k_c$, given in (10). The algorithms used have been shown to be asymptotically exhaustive by Stark (1987).

CHAOS AND ERGODIC THEORY

According to one point of view, expressed by Laplace, dynamical systems like the Solar System are completely deterministic, so probability theory can have no relevance. But this point of view requires a God-like omniscience in being able to determine initial conditions exactly. This requires an infinite number of digits and is beyond the capacity of anybody or anything of finite size, including the observable Universe (Ford 1983).

In reality measurement is only able to determine the state of a classical system to a finite number of digits, and even this determination is subject to errors, without quantum mechanics, and whether the determination is made by human or machine. Such measurements limit the known or recorded motion to a range of possible orbits.

For realistic mixed systems, the theory of such incomplete conditions is complicated and the theory has not been worked out, but for the Anosov systems, which are chaotic almost everywhere, the ergodic theorists have introduced symbolic dynamics, which shows precisely how probabilities enter into deterministic dynamics (Arnol'd & Avez 1968; Moser 1973; Shields 1973; Ornstein 1974; Sinai 1976).

Symbolic dynamics relates three things:

sequences of symbols;

orbits of a dynamical system;

partitions of its phase space.

The partitions are fine grained, but they are generated by the dynamics itself from simple very coarse-grained partitions.

Kolmogorov and Sinai introduced a rate of entropy production, known as the K–S entropy, which measures the amount of chaos in purely chaotic (Anosov) system. The concept was taken from Shannon's theory for the symbol sequences of communication theory. For such systems the K–S entropy is equal to the Liapounov exponent, which measures the mean divergence of neighbouring orbits.

The relation has been used very effectively for mixed systems, as Liapounov exponents are easier to calculate than the K–S entropy, but there is no complete theory for these systems.

Thus integrable systems, for which the motion is purely regular, and Anosov systems, for which the motion is chaotic almost everywhere, are well understood, whereas most real systems are mixed, and there is no comparable theory for them.

Mixed systems

Despite the absence of symbolic dynamics for mixed systems, numerical experiment backed by dynamical theory has helped us to understand their chaotic behaviour.

For problems of confinement of particles, and of atoms in molecules, the destruction of invariant tori is crucial. For area preserving maps, representing conservative systems of two degrees of freedom, an invariant curve in the phase space forms an impassable barrier to the chaotic motion of a phase point.

When a torus is destroyed, for values of k greater than k_c, an invariant cantor set known as a cantorus remains (Aubry 1978; Percival 1979b). Cantori have been proved to exist for area-preserving twist maps such as the standard map by Aubry and Le Daeron (1983) and by Mather (1982a). An approximate cantorus for the standard map is illustrated in figure 6. Variational principles for invariant sets provide both insight and proofs in this theory (Percival 1979*a*, *b*; Klein & Lee 1979; Mather 1982*b*).

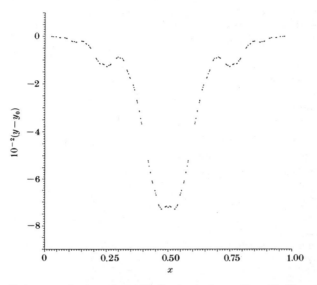

FIGURE 6. An approximate cantorus for the standard map (from MacKay *et al.* 1984).

As one might expect, cantori impede the flux of phase points in the chaotic regions of mixed systems, even though they do not prevent it, and thus produce partial confinement, which is most effective when the value of k is close to the critical value. The flux of phase points is measured in terms of areas obtained from differences in action between orbits of various types, including orbits on cantori, and orbits that tend to cantorus orbits or periodic orbits in the limit of infinitely long times in the past or future (MacKay *et al.* 1984, 1987; BenSimon & Kadanoff 1984).

For systems of more than two degrees of freedom, invariant tori and cantori do not have sufficiently high dimension to confine the motion in the chaotic regions of most mixed systems. Chaotic motion, known as Arnol'd diffusion, can take place in an 'Arnol'd web' throughout the phase space, subject to conservation laws, even for arbitrarily small perturbations of an integrable system (Arnol'd 1964; Lichtenberg & Lieberman 1983).

The long-term chaotic dynamics of mixed systems is considerably complicated by the presence of island chains of all orders (Meiss 1986), and by the stickiness of the boundary between the chaotic and regular regions of phase space. A satisfactory general theory does not exist.

Applications

The new view of dynamics has developed in parallel with its applications. These include the classical statistical mechanics of systems with few degrees of freedom such as stars in galaxies and atoms in molecules, the confinement of charged particles in particle accelerators, storage rings and laboratory plasmas and also the semiclassical limit of quantal systems.

Only relatively recently have the new ideas had a significant impact on the dynamics of the Solar System, the original application which stimulated the work of Poincaré. Now that their relevance to the Solar System has been appreciated, there has been a wealth of new results (Wisdom, this symposium and references therein; Dermott & Murray 1983; Murray & Fox 1984; Milani *et al.* 1986; Chirikov & Vecheslavov 1986).

Conclusions

The complete change of viewpoint described earlier can now be summarized, as in table 1, showing the old ideas, and how they have been modified in the last few decades.

Table 1. Changing viewpoints in hamiltonian dynamics

old view	new view
integrability	invariant sets
stability and instability of orbits	regular and chaotic motion
variational principles for orbits	variational principles for invariant sets
convergence of perturbation series	K.A.M. theorem and analyticity in complex domains
continuity	fractals and renormalization
analysis	geometry, computers and number theory
determinism	probability

I should like to thank V. I. Arnol'd, R. S. MacKay, C. Murray and F. Vivaldi for helpful communications, and the SERC for research support.

References

Arnol'd, V. I. 1961 *Izv. Akad. Nauk SSSR ser. Mat.* **25**, 21.
Arnol'd, V. I. 1963 *Usp. mat. Nauk* **18**, 13. (Transl. *Russian Mathematical Surveys* **18**, 91 (1963).)
Arnol'd, V. I. 1964 *Dokl. Akad. Nauk USSR* **156**, 9.
Arnol'd, V. I. & Avez, A. 1968 *Ergodic problems of classical mechanics.* New York: W. A. Benjamin.
Aubry, S. 1978 In *Solitons and condensed matter physics* (ed. A. R. Bishop & T. Schneider) (Springer Series in Solid State Sciences no. 8), pp. 264–277. New York: Springer.
Aubry, S. 1983 *Physica* D **7**, 240.
Aubry, S. & Le Daeron, P. Y. 1983 *Physica* D **8**, 381.
Aubry, S., Le Daeron, P. Y. & André, G. 1982 Classical ground states of a 1-D model for incommensurate structures. Preprint C. E. N. Saclay.
BenSimon, D. & Kadanoff, L. P. 1984 *Physica* D **13**, 82.
Birkhoff, G. D. 1920 *Acta Math.* **43**, 1.
Birkhoff, G. D. 1932 *Bull. Soc. Math. Fr.* **60**, 1.
Celletti, A., Falcolini, C. & Porzio, A. 1987 *Annls Inst. Henri Poincaré.*

Chirikov, B. V. 1969 Research concerning the theory of nonlinear resonance and stochasticity. Nuclear physics section of the Siberian Academy of Sciences, report 267. (In Russian.)
Chirikov, B. V. 1971 CERN translation 71-40, Geneva.
Chirikov, B. V. 1979 *Physics Rep.* **52**, 263–379.
Chirikov, B. V. & Vecheslavov, V. V. 1986 Chaotic dynamics of Comet Halley. Preprint 86–184. Institute of Nuclear Physics, Novosibirsk.
Dermott, S. F. & Murray, C. D. 1983 *Nature, Lond.* **301**, 201–205.
Ford, J. 1983 *Physics Today* **36**, 40–47.
Froeschlé, C. 1970 *Astron. Astrophys.* **9**, 15–23.
Goward, F. K. 1953 In *Lectures on the theory and design of an alternating gradient proton synchrotron*, p. 19. Geneva: CERN.
Greene, J. M. 1979 *J. math. Phys.* **20**, 1183.
Greene, J. M. & Percival, I. C. 1981 *Physica D* **3**, 530.
Hénon, M. 1966 *Bull. astr., Paris* **1**, 49–66.
Herman, M. R. 1983 *Astérisque*, nos 103–104.
Herman, M. R. 1986 *Astérisque*, no. 144.
Hine, M. G. N. 1953 In *Lectures on the theory and design of an alternating gradient proton synchrotron*, p. 69. Geneva: CERN.
Klein, A. & Lee, C.-T. 1979 *J. math. Phys.* **20**, 572–578.
Kolmogorov, A. N. 1954 *Dokl. Akad. Nauk SSSR* **98**, 527.
Kolmogorov, A. N. 1957 In *Proc. 7th International Congress of Mathematicians, 1954* (ed. J. C. H. Gerretson & J. de Groot), vol. 1, p. 315. (In Russian.) (Transl. R. Abraham *Foundations of mechanics*, appendix D. New York: W. A. Benjamin (1967).)
Lichtenberg, A. J. & Lieberman, M. A. 1983 *Regular and stochastic motion.* (Applied Mathematical Sciences 38.) New York: Springer.
MacKay, R. S., Meiss, J. D. & Percival, I. C. 1984 *Physica D* **13**, 55.
MacKay, R. S., Meiss, J. D. & Percival, I. C. 1987 *Physica D.* (In the press.)
MacKay, R. S. & Percival, I. C. 1985 *Communs math. Phys.* **98**, 469–512.
Mather, J. N. 1982*a* *Ergodic Theory Dynam. Syst.* **2**, 397.
Mather, J. N. 1982*b* *Topology* **21**, 457–467.
Mather, J. N. 1982*c* *Communs math. Phys.* **86**, 465–473.
Mather, J. N. 1984 *Ergodic theory Dynam. Syst.* **4**, 301.
Meiss, J. D. 1986 *Phys. Rev. A* **34**, 2375–2383.
Milani, A., Nobili, A. M., Fox, K. & Carpino, M. 1986 *Nature, Lond.* **319**, 386.
Moser, J. 1962 *Nachr. Akad. Wiss. Göttingen* **1**, 1.
Moser, J. 1973 Stable and random motion in dynamical systems. Princeton University Press.
Moser, J. 1974 In *The stability of the Solar System* (ed. Y. Kozai), pp. 1–9. International Astronomical Union.
Newman, R. A. P. C. & Percival I. C. 1983 *Physica D* **6**, 249.
Murray, C. D. & Fox, K. 1984 *Icarus* **59**, 221–233.
Ornstein, D. S. 1974 Ergodic theory, randomness and dynamical systems. Yale University Press Mathematical Monographs.
Percival, I. C. 1979*a* *J. Phys. A* **12** L57–L60.
Percival, I. C. 1979*b* In *Nonlinear dynamics and the beam–beam interaction* (ed. M. Month & J. C. Herrera), vol. 57, pp. 302–310. American Institute of Physics Conference Proceedings.
Percival, I. C. 1982 *Physica D* **6**, 67–77.
Poincaré, H. 1899 *Les méthodes nouvelles de la méchanique céleste*, vols 1, 2 and 3. Paris: Gauthier-Villars.
Poincaré, H. 1967 *New methods of celestial mechanics*, vols 1, 2 and 3, NASA Technical Translation TT F-450.
Poincaré, H. 1896 *C. r. hebd. Séanc. Acad. Sci., Paris* **22**, 497–499.
Shields, P. 1973 *Theory of Bernoulli Shifts*. University of Chicago Press.
Sinai, Ya. G. 1976 *Introduction to ergodic theory*. Princeton University Press.
Symon, K. R. & Sessler, A. M 1956 In *Proc. CERN Symposium on High Energy Accelerators and Pion Physics, Geneva*, pp. 44–58.
Stark, J. 1987 *Communs Math. Phys.* (Submitted.)
Taylor, J. B. 1969 Culham Progress Report CLM-PR-12.

Particle confinement and adiabatic invariance

By B. V. Chirikov

Institute of Nuclear Physics, Prospekt Nauky 11, *6300 Novosibirsk* 90, *U.S.S.R.*

The general problem of adiabatic invariance is discussed by using, as an example, the particle motion in a mirror magnetic trap. A new, resonant, adiabaticity parameter is introduced, which allows the reduction of the axisymmetric problem to the standard map, and which provides a sharp estimate for the chaos (instability) border in magnetic traps. Peculiarities of the particle diffusion out of the trap, due to the presence of the chaos border, are considered. The mechanism of slow Arnol'd diffusion in a non-axisymmetric field is explained, and some rough estimates are given, including the impact of weak noise (gas scattering).

1. Introduction

The problem of controlled nuclear fusion, aimed at the peaceful use of thermo-nuclear power, has stimulated a broad range of fundamental research in plasma physics. Among other problems, particle dynamics in electromagnetic fields has been intensively studied. On one hand, these fields are supposed to confine particles for a fairly long time within a bounded domain of space (the so-called 'magnetic traps'). On the other, they are expected to heat the same particles up to a very high thermonuclear temperature.

In these studies physicists have come across the first examples of a rather peculiar phenomenon, called *dynamical chaos*, that is a random (unpredictable) motion of a completely deterministic mechanical system whose equations of motion did not contain any random parameters or any noise (see, for example, Chirikov 1960; Lichtenberg & Lieberman 1983; Chirikov 1984; Zaslavsky 1985). Still earlier, similar phenomena were observed in numerical experiments (Goward 1953; Hine 1953) on particle dynamics in an accelerator (also a 'magnetic trap').

Apparently the first who actually dealt with dynamical chaos was English psychologist and anthropologist, Galton. In 1889 (almost a century ago!) he had developed his famous device, the Galton Board, to demonstrate statistical laws as well as statistical methods for studying mass phenomena, the methods he first introduced into psychology (Galton 1889). Galton himself had, apparently, no interest in the dynamics of his fairly simple mechanical model which has been termed later on as 'Lorentz's gas', and which nowadays is intensively studied, among others, by the Soviet mathematician Sinai and his disciples. Only 17 years later Poincaré conjectured that a highly irregular, chaotic motion in such systems was due to a strong local instability of the trajectories. It took more than half a century to develop this fairly simple physical insight into a rigorous theory. Now motion instability or, in modern language, a positive (non-zero) Kolmogorov–Sinai entropy is the most efficient criterion for dynamical chaos (Alekseev *et al.* 1981), although still not commonly accepted.

Back to plasma physics, I shall mention also the problems concerning the 'dynamics' of a magnetic line itself (Gelfand *et al.* 1963; Rosenbluth *et al.* 1966; Filonenko *et al.* 1967). These lines may be chaotic as well (a 'braided' magnetic field), which results in a considerable increase of the plasma transverse thermal conductivity. Particle diffusion in such a field (Rechester *et al.* 1978; Kadomtsev *et al.* 1978) provides a striking example of time-reversible chaos. Consider a toroidal magnetic trap (say, tokamak) and write the diffusion of magnetic lines as $\Delta r_B \propto s^{\frac{1}{2}}$ with s the distance along a line. Assuming that particle follows a certain particular magnetic line and neglecting the scattering completely we arrive at the standard diffusion estimate $\Delta r_p \propto t^{\frac{1}{2}}$. Now, take the scattering into account, still neglecting all the finite Larmor radius effects. Then, the particle displacement along a line grows as $s \propto t^{\frac{1}{2}}$ only, because of random changes of the sign of its velocity as a result of scattering. Hence the transverse diffusion drops down to $\Delta r_p \propto t^{\frac{1}{4}}$ (Rechester *et al.* 1978). Of course, the particle is going to leave its 'own' chaotic magnetic line eventually, yet the plasma thermal conductivity considerably decreases. It is an important practical consequence of the time reversibility of dynamical chaos.

2. BUDKER'S PROBLEM

Now we are going to discuss in some detail a typical dynamical problem in the plasma theory, the particle motion in a mirror magnetic trap. Those devices are now called 'open' (for the magnetic lines, and we hope not for the plasma!). The problem was formulated in 1954 by Soviet physicist Budker who proposed, independently from American physicists, to make use of such a trap for controlled nuclear fusion (see Budker 1982). Precisely this problem was the beginning of the study of dynamical chaos in our institute, at that time in the depths of the Kurchatov Institute for Atomic Energy.

Particle confinement in an open trap is provided by the conservation of an adiabatic invariant, the magnetic moment of a particle, which is known, however, to be an approximate motion integral only. Thus, Budker's problem turned out to be a particular case of the general fundamental problem of *adiabatic invariance*, that is the invariance of the action variables under the special, adiabatic perturbation.

Previously, *pulsed*, or *single*, adiabatic perturbations, which are effective over a finite time interval only, had been mainly studied. As is well known, the principal condition for adiabatic invariance in this case is perturbation *slowness* as compared with the characteristic frequencies of a system. In the class of analytical functions the violation of adiabaticity is asymptotically (for $t \rightarrow \pm \infty$) exponentially small in the slowness parameter, and hence unimportant as a rule.

In Budker's problem we encounter a rather different situation when the adiabatic perturbation, due to a particle longitudinal libration, is a *stationary*, or *multiple* one. The smallness of the libration frequency as compared with the Larmor frequency (the slowness of adiabatic perturbation) does no longer guarantee the conservation of the adiabatic invariant because its small changes over each libration period may now accumulate. Here, as in many other problems in the theory of oscillations, the resonances (linear or nonlinear) between the particle gyration and libration are of primary importance.

A vague idea about the role of resonances in the problem of adiabatic invariance arose at the dawn of the quantum mechanics, where the action variables are of a special importance. Yet this question was only properly formulated and answered in 1928, for linear oscillations, by Soviet physicists Andronov, Leontovich and Mandelstam (see Mandelstam 1948). Remarkably, it proved to be sufficient simply to carefully look at the well-known Mathieu's equation and its solutions from the physical viewpoint. Indeed, the stop bands of parametric resonance do exist here for arbitrary small values of the slowness parameter

$$\epsilon = \Omega/2\omega_0 \qquad (1)$$

near $\epsilon = 1/n$, where $n \neq 0$ is any integer, ω_0 the unperturbed frequency of a linear oscillator, and Ω the frequency of the harmonic parametric perturbation. Thus, in the present case, the necessary condition for adiabatic invariance is the *lack of resonances* in perturbation. Its slowness just helps in that the resonance width $(\Delta\omega_0)$ as well as the maximal increment (γ) of the instability (in action) drop exponentially in ϵ:

$$\Delta\omega_0 \approx \gamma \approx \tfrac{1}{3}\Omega(\tfrac{1}{8}e^2 g)^{1/\epsilon}. \qquad (2)$$

Here $e = 2.71...$, and g is dimensionless perturbation amplitude: $\omega^2(t) = \omega_0^2(1 - g\cos\Omega t)$ (see Chirikov 1984; 1986 for details).

The problem becomes much more complicated for the nonlinear oscillations of particle in a magnetic trap. First, the resonances now depend not only on the trap parameters but also on the motion initial conditions, which determine the oscillation frequencies. Moreover, a single resonance does not produce any instability at all because of the stabilization of resonant perturbation by non-linearity. Yet, the interaction of several nonlinear resonances (if only two) results in a rather peculiar instability, dynamical chaos (see, for example, Lichtenberg & Lieberman 1983; Zaslavsky 1985).

3. RESONANT PERTURBATION THEORY FOR ADIABATIC PROBLEMS

Consider, as a particular example, the axisymmetric trap whose magnetic field in some axis neighbourhood depends on the coordinate s along a magnetic line as

$$\omega(s) = \tfrac{1}{2}\omega_0[(\lambda+1) - (\lambda-1)\cos\pi s/L]. \qquad (3)$$

Here ω_0 is the minimal magnetic field, and λ the mirror ratio. This model also describes a multimirror ($2L$ being the mirror spacing), or 'corrugated' magnetic field.

An important step from the beginning is the choice of unperturbed system. We require the magnetic moment to be constant,

$$\mu = v_\perp^2/2\omega \equiv \text{const.}, \qquad (4)$$

where $v_\perp$ is the transverse component of particle velocity. That choice corresponds to the well-known Born–Oppenheimer approximation in quantum mechanics. The unperturbed hamiltonian becomes

$$H^0(p, s) = \tfrac{1}{2}v^2 = \tfrac{1}{2}p^2 + \mu\omega(s), \qquad (5)$$

where $p = \dot{s}$ is the momentum. We use relativistic units $e = c = m(1-v^2)^{-\frac{1}{2}} = 1$. All the relations hold for an arbitrary particle velocity even though some of them have a 'non-relativistic' appearance.

Below we restrict ourselves for simplicity to two limiting cases

$$1/\lambda \ll \sin^2\theta_0 \ll 1, \tag{6a}$$

$$|1/\lambda - \sin^2\theta_0| \ll 1/\lambda. \tag{6b}$$

Here θ_0 is the pitch-angle at the field minimum: $\mu = v^2 \sin^2\theta_0/2\omega_0$. Case (a) corresponds to a small (harmonic) libration in the 'potential'

$$U(s) \approx \omega_0(1 + s^2/s_{\mathrm{p}}^2); \quad s_{\mathrm{p}} = 2L/\pi\sqrt{(\lambda-1)} \ll L, \tag{7}$$

where s_{p} characterizes the spatial field scale. For (b) the motion is close to separatrix $H^0 = H_{\mathrm{s}}^0 = \mu\lambda\omega_0$, or the loss cone $\theta_0 = \theta_{\mathrm{s}}$, $\sin\theta_{\mathrm{s}} = \lambda^{-\frac{1}{2}} \ll 1$, which separates the trapped ($\Delta\theta_0 = \theta_0 - \theta_{\mathrm{s}} > 0$) and untrapped trajectories. It is convenient to describe the distance from the separatrix by the dimensionless quantity

$$w(\mu) = \frac{2(H^0 - H_{\mathrm{s}}^0)}{\mu(\lambda-1)\omega_0} \approx -4\sqrt{\lambda}\Delta\theta_0; \quad |w| \ll 1. \tag{8}$$

The unperturbed frequencies are:

$$\Omega(\mu) = \frac{\partial H^0}{\partial I} \approx \frac{\sqrt{(2\mu\omega_0)}}{s_{\mathrm{p}}} \equiv \Omega_0(\mu),$$

$$\langle\omega(\mu, I)\rangle = \frac{\partial H^0}{\partial\mu} \approx \omega_0 + \frac{\pi}{2\sqrt{2}}\sqrt{\left(\frac{\lambda\omega_0}{\mu}\right)}\frac{I}{L} \approx \frac{\omega_0}{2\sin^2\theta_0}, \tag{9a}$$

$$\Omega(\mu) \approx \frac{\pi\Omega_0}{\Lambda(\mu)}; \quad \langle\omega(\mu)\rangle \approx \lambda\omega_0\left(1 - \frac{2}{\Lambda(\mu)}\right) \approx \lambda\omega_0. \tag{9b}$$

Here, $\Lambda(\mu) = \ln(32/|w|)$, I is longitudinal action, and $\langle\omega\rangle$ the Larmor gyrofrequency averaged over the libration period. Define the slowness parameter of the adiabatic perturbation as

$$\epsilon(\mu, I) = \frac{\Omega}{\langle\omega\rangle} \approx \frac{2\mu}{I} \approx \pi\frac{\rho_{\mathrm{m}}}{L}\sqrt{\lambda}\sin^3\theta_0, \tag{10a}$$

$$\epsilon(\mu) \approx \frac{\pi\Omega_0}{\lambda\omega_0(\Lambda-2)} \approx \frac{\pi^2}{2}\frac{\rho_{\mathrm{m}}}{L}\frac{\sqrt{\lambda}}{\Lambda}\sin\theta_0, \tag{10b}$$

where $\rho_{\mathrm{m}} = v/\omega_0$ is the maximal Larmor radius. Notice that even for (a), *harmonic* libration (7), the latter is essentially *nonlinear* because its frequency $(9a)$ as well as the frequency ratio $(10a)$ depend on action variables μ, I. Both actions are related by the energy conservation $H^0 = \mathrm{const.}$, whence $I \approx LH^0/\sqrt{(2\omega_0\mu)}$ for (a).

A specific difficulty of stationary adiabatic problems in the theory of nonlinear oscillations is that one cannot directly use here the powerful asymptotic methods in the perturbation theory, particularly, a fairly simple and efficient averaging method. This can be seen from estimate (2) for the linear problem. Even though the dependence on the perturbation amplitude is a power law, which can be obtained via a standard asymptotic series, the expansion in the small slowness parameter ϵ fails. On the other hand, at large perturbation amplitudes, which are typical for adiabatic processes, ϵ is the only small parameter. An example is the model under consideration (3) for $\lambda \gg 1$. The difficulty is still worse because for stationary oscillations the change in action variables consists of two quite different parts:

(i) the *quasiperiodic* variation, for instance, $\Delta\mu_q \sim \epsilon$, which is relatively big but non-cumulative and, hence, unimportant after all;

(ii) the *resonant* variation, $\Delta\mu_r \sim e^{-1/\epsilon}$ (over a libration period, see (12) below), which just results in motion instability and particle losses even though it is rather small compared with (i).

To overcome this difficulty, introduce a new 'good' adiabaticity parameter which, first, would describe the resonant cumulating variation $\Delta\mu_r$ only, and, second, would include the non-analytical exponential $\Delta\mu_r \sim e^{-1/\epsilon}$ from the beginning. Both objectives are naturally realized by the construction of a map over the libration period. Because of the symmetry of model (3) in respect to the field minimum, the map over half a period is sufficient. At any resonance ($m\langle\omega\rangle = 2n\Omega$), the quasiperiodic variations $\Delta\mu_q$ can be completely excluded, hence the term '*resonant perturbation theory*'. Actually, such an approach had been used already by Chirikov (1960) and was further developed later on (Chirikov 1978; see also Chirikov 1979, 1984).

To realize this approach one needs, first of all, to calculate $\Delta\mu_r$ by some direct integration. Curiously, this problem has been solved initially for the quantum equations of motion with subsequent transition to the classical mechanics (Dykhne *et al.* 1961), an amusing zigzag of cognition! A simple classical procedure for evaluating $\Delta\mu_r$ has been developed and applied by Hastie *et al.* (1968).

For the field (3), $\Delta\mu_r$ is exponentially small because of the (per period) singularity $\omega(s) = 0$ at $s = is_p$ (7). Because $s_p \ll L$ for $\lambda \gg 1$ ($\theta_0 \ll 1$) the change in μ occurs in a small neighbourhood of the field minimum. Introducing an new dimensionless variable

$$P = \sqrt{(\mu/\mu_{max})} = \sin\theta_0 \approx \theta_0 \ll 1 \qquad (11)$$

we obtain in both cases ((a) and (b)) (Chirikov 1984)

$$\xi \equiv (\Delta P)_{max} \approx \frac{9\pi^2}{32} \frac{r_0\sqrt{\lambda}}{L} \frac{e^{-1/\epsilon_B}}{\epsilon_B}; \quad \epsilon_B = \frac{3\pi}{4} \frac{\rho_m\sqrt{\lambda}}{L} \ll 1 \qquad (12)$$

for the amplitude of the variation of P, which we choose as a new 'good' adiabaticity parameter ξ. Here r_0 is the distance of the Larmor centre from the symmetry axis at a field minimum; $\epsilon_B \sim \epsilon$ the field 'smoothness' parameter (cf. (10)), and the last inequality is the applicability condition for (12).

4. THE STANDARD MAP

The resonant change in P depends on the Larmor phase ϕ at the field minimum: $\Delta P = \xi \cdot \sin\phi$. In turn, the ϕ variation is determined by the frequency ratio (10), or by the parameter of perturbation slowness: $\Delta\phi = \pi/\epsilon(P)$. This leads to a map $(P, \phi) \rightarrow (\bar{P}, \bar{\phi})$ over half a libration period

$$\bar{P} = P + \xi\sin\phi; \quad \bar{\phi} = \phi + \pi/\epsilon(\bar{P}), \qquad (13)$$

which describes a long-term particle motion.

The cumulative non-adiabatic variation of μ is related to resonances at $P = P_n$ where $\epsilon(P_n) = \frac{1}{2}n$ with any integer n. The rate of accumulation is determined by the new parameter ξ, which may be called therefore the *resonant adiabaticity parameter* as well as the whole perturbation theory based upon it (§3). Notice that high harmonics of the libration frequency, which are required for the resonances at large n, are present even in the case (a) of harmonic oscillations due to the libration-modulated Larmor frequency.

Because ξ does include the principal non-asymptotic effect of adiabatic perturbation, one can use any standard asymptotic expansion in ξ, particularly, the efficient averaging method (Chirikov 1960, 1978, 1979, 1984).

Map (13) can be simplified still further by linearizing the second equation in P. Introducing a new variable $p = G(P) + G'(P)\Delta P$ with $G(P) = \pi/\epsilon$ we arrive at the so-called standard map

$$\bar{p} = p + K\sin\phi; \quad \bar{\phi} = \phi + \bar{p}, \tag{14}$$

which describes the local (in P) dynamics of model (13). These depend on the only parameter

$$K(P) = \xi|G'(P)| \approx \frac{81\pi^3}{128} \frac{r_0 \sqrt{\lambda}}{L} \frac{e^{-1/\epsilon_B}}{\epsilon_B^2 P^4}, \tag{15a}$$

$$K(P) \approx \frac{27\pi^2}{64} \frac{r_0 \lambda^2}{L} \frac{e^{-1/\epsilon_B}}{\epsilon_B^2 |\Delta P|}. \tag{15b}$$

Here $\Delta P = P - P_s \approx \Delta\theta_0; P_s = \sin\theta_s$. We shall call K the *stability parameter* because there exists a critical value K_c separating bounded motion in $p(K \leqslant K_c)$ from unbounded motion $(K > K_c)$. The latter just means the instability (non-adiabaticity) of particle motion that brings the particle onto the loss cone at $P = P_s = \lambda^{-\frac{1}{2}} \ll 1$. In terms of nonlinear resonances the critical K is determined by their 'overlapping' which results in an unbounded 'wandering' of trajectory in p (Chirikov 1979). On the other hand, the Kolmogorov–Arnol'd–Moser (K.A.M.) theory does rigorously guarantee (for $K \to 0$) the boundedness of p and μ oscillation, which results in the eternal confinement of a particle in the trap (Arnol'd 1963). In this sense the adiabatic invariant becomes an exact one. Notice that for some initial conditions the eternal oscillation of a particle is, nevertheless, chaotic. Thus we encounter the curious phenomenon of *chaotic adiabaticity*. On the other hand, there are domains of regular bounded oscillations for $K > K_c$ as well.

The precise K_c value is obtained in the theory of critical phenomena, which is a recent development of the K.A.M. theory. According to Greene (1979) and to MacKay & Percival (1985) the critical K lies within the interval

$$0.9716... \leqslant K_c < \tfrac{63}{64} = 0.9843.... \tag{16}$$

Such an uncertainty is of no importance for applications, of course, yet it leaves open some interesting questions in the theory of the critical structure (Chirikov & Shepelyansky 1986).

The elusively simple model (14) has turned out to be very popular in the studies on nonlinear and chaotic dynamics. Besides, as was shown above, the standard map has a direct bearing on some real physical problems. A new example is the dynamics of a comet under Jupiter's perturbation (Petrosky 1986).

Among the early researchers on model (14) was British physicist J. B. Taylor.

To the best of my knowledge, the standard map first appeared in the problem on electron dynamics in a new relativistic accelerator, the microtron, invented by Soviet physicist Veksler (1944). This was studied by Kolmensky (1960) and many others (see, for example, Kapitsa & Melekhin 1969). In all these papers the case of a stable regular acceleration only was considered. The main microtron domain corresponds to the standard map parameter K as follows

$$K = \frac{2\pi\omega V}{cB} = \frac{2\pi}{\sin\phi_s} \approx 6.59, \tag{17}$$

where ω, V, ϕ_s are the frequency, amplitude, and equilibrium phase of the accelerating voltage respectively, and where B is the magnetic field strength. The value (17) is much greater than the critical one. As a result, the regular acceleration domain is only about 1 % of the phase plane, even for the optimal $\sin\phi_s \approx 0.95$. The other initial conditions give rise to chaotic electron motion, that is the microtron turns into a 'stochatron', upon some necessary changes in its design, of course. Such a chaotic electron acceleration has been observed recently in a 'plasma microtron' (Vaskov *et al.* 1984).

5. THE CHAOS BORDER AND STATISTICAL ANOMALIES

Back to Budker's problem we see from (15) that there is always a *chaos border* in the magnetic trap determined by the condition $K = K_c \approx 1$. In velocity space the border is a cone with vertex angle $\theta_b \approx P_b$, where

$$\frac{1}{\sqrt{\lambda}} \ll \theta_b \approx 2.1 \left(\frac{r_0}{4}\right)^{\frac{1}{4}} \lambda^{\frac{1}{8}} \frac{e^{-\frac{1}{4}\epsilon_B}}{\epsilon_B^{\frac{1}{2}}} \ll 1, \tag{18a}$$

$$|\Delta\theta_b| \approx 4.2 \frac{r_0\lambda^2}{4} \frac{e^{-1/\epsilon_B}}{\epsilon_B^2} \ll \frac{1}{\sqrt{\lambda}}. \tag{18b}$$

The inequalities recall the applicability conditions for simplifying assumptions of §3. One may say that the chaos border widens the adiabatic loss cone $\theta_s \approx \lambda^{-\frac{1}{2}}$.

A very intricate critical structure near the chaos border consists of alternating components of regular and chaotic motions whose spatial (in θ) scales indefinitely decrease towards the border while the temporal ones increase without limit. This leads to a long-time 'sticking' of a chaotic trajectory, and, hence, to a sharp drop of the diffusion rate. As a result, the statistical relaxation proceeds abnormally slowly according to some power, rather than exponential, law.

If, for example, the particles fill up all the chaotic component connected with the adiabatic loss cone homogeneously, the number of still confining particles $N(t)$ first decays exponentially in time because of an ordinary relatively fast diffusion off the chaos border. However, as soon as the relaxation process approaches the border its rate drops and, according to numerical experiments (Chirikov & Shepelyansky 1981; Karney 1983), proceeds approximately as follows:

$$N(t) \propto t^{-\frac{1}{2}}. \tag{19}$$

The time correlation in the chaotic motion with a border decays in the same way.

However, if the particles are injected near the adiabatic loss cone, i.e. relatively far from the chaos border, than $N(t)$ decays according to (19) *from the beginning* but for a different reason, namely homogeneous diffusion inside the trap. In this case $N(t)$ is proportional to the integral probability of Poincaré's recurrences into the initial state, at the loss cone. For different models this initial stage of the process was observed in some numerical experiments (Channon & Lebowitz 1980; Yamaguchi & Sakai 1986). However, as soon as the diffusion reaches the chaos border the decay proceeds as (Chirikov & Shepelyansky 1981)

$$N(t) \propto t^{-\frac{3}{2}}. \tag{20}$$

The latter régime was apparently observed by Indian physicists in a laboratory experiment on electron dynamics in a magnetic trap (Bora *et al.* 1980). The empirical value of the exponent in (20) was approximately -1.3 (Chirikov 1984).

A power-law correlation decay $C(\tau) \propto \tau^{-p}$ with the exponent $p < 1$ (19) qualitatively transforms the diffusion related to that correlation. Formally, the diffusion rate becomes infinite. Actually, it means that the dispersion σ^2 of distribution function grows abnormally fast (Chirikov & Shepelyansky 1984):

$$\sigma^2(t) \propto t^{2-p} \approx t^{\frac{3}{2}}. \tag{21}$$

This occurs, for instance, in the chaotic component of the standard map in the presence of a domain of regular microtron acceleration (17) with the chaos border around, that is when the microtron behaves as a stochatron. In numerical experiments (Karney *et al.* 1982) the empirical 'diffusion rate' increased by a factor of about 100, even though the relative area of the regular microtron acceleration was as small as $A_s \approx 0.02$. The theory of critical phenomena leads to the estimate

$$\sigma^2 \approx \alpha A_s \tfrac{1}{2} K^2 t^{\frac{3}{2}}, \tag{22}$$

where $\alpha \approx 0.5$ from the numerical data (Karney *et al.* 1982).

6. Arnol'd diffusion and universal non-adiabaticity

Above we considered the axisymmetric magnetic trap where the particle motion is essentially two dimensional (two degrees of freedom). Any asymmetry of the magnetic field, making the motion three dimensional, greatly complicates the particle dynamics so that only rough estimates can be derived. In what follows we are going to discuss the three typical cases.

(a) A strong asymmetry, the multiplet overlap

Because of a particle drift with frequency Ω_d each of resonances $\langle \omega \rangle = 2n\Omega$ (§4) splits into a multiplet of m subresonances with spacing Ω_d and with

$$m \approx \frac{\Delta\langle\omega\rangle + 2n\Delta\Omega}{\Omega_d} \approx \left[\frac{\Delta\langle\omega\rangle}{\langle\omega\rangle} + \frac{\Delta\Omega}{\Omega}\right]\frac{\langle\omega\rangle}{\Omega_d} \equiv \nu\frac{\langle\omega\rangle}{\Omega_d}, \tag{23}$$

where $\Delta\langle\omega\rangle$, $\Delta\Omega$ are the full widths of frequency modulation on a *drift surface* (Chirikov 1978). Under condition $\nu \gtrsim 2\Omega/\langle\omega\rangle \approx 1/n \ll 1$ the neighbouring multiplets do overlap, and a usual global chaos sets in. The critical K value considerably diminishes

$$K_c(\nu) \approx \sqrt{(\pi m)} \, (\Omega_d/2\Omega)^2 \sim \epsilon\sqrt{\nu} \gtrsim \epsilon^{\frac{3}{2}} \ll 1. \tag{24}$$

In the last estimate $\Omega_d/\Omega \sim \epsilon = \Omega/\langle\omega\rangle$ is assumed, and the lower bound for $K_c(\nu)$ is determined by the multiplet overlap. For $K > K_c(\nu)$, the diffusion rate is about that at $\nu = 0(K > 1)$ as it depends on the spectral density of perturbation.

(b) A moderate asymmetry, the modulational diffusion

Let

$$\epsilon^2 \sim \Omega_d/\langle\omega\rangle \lesssim \nu \lesssim 1/n \sim \epsilon, \tag{25}$$

then multiplets do not overlap, yet they exist as $m \gtrsim 1$ (the left inequality). Under condition $K \gtrsim K_c(\nu)$ (24) the resonances within a multiplet do overlap and form a solid chaotic layer along a resonance $\langle\omega\rangle = 2n\Omega$, the layer width being $\Delta\omega \approx m\Omega_d \approx \nu\langle\omega\rangle$. In the absence of multiplet overlap the diffusion can still go along resonance $\langle\omega\rangle = 2n\Omega$ (for a simple model of such a *modulational diffusion* see, for example, Lichtenberg & Lieberman 1983; Vivaldi 1984).

According to (10a), modulational diffusion proceeds in both θ_0 and r_0 while

$$\frac{\langle\omega(r_0)\rangle}{\sin^3\theta_0} \approx \text{const.} \tag{26}$$

Because the magnetic field goes down with r_0 any decrease in θ_0 leads to an increase in r_0. Notice that in the axisymmetric trap the radial motion is forbidden by the conservation of angular momentum.

In our case the non-axisymmetric perturbation is a high-frequency one that is the detuning from the nearest resonance $\delta\omega \sim \Omega \gg \Delta\omega \approx \nu\langle\omega\rangle \sim \nu\Omega/\epsilon$ is much larger than the multiplet width. It is also adiabatic as the inverse frequency ratio is small. Hence, the effect of this perturbation is exponentially small too, and the corresponding resonant adiabaticity parameter ξ_M can be roughly estimated as

$$\xi_M \sim \nu \exp\left(-2\pi\frac{|\delta\omega|}{\Delta\omega}\right) \sim \nu\, e^{-a\epsilon/\nu} \gtrsim \epsilon^2\, e^{-a/\nu}; \quad a \sim 1. \tag{27}$$

For a more accurate evaluation see Vivaldi (1984); Chirikov *et al.* (1985).

With initial condition inside any multiplet the diffusion occurs, and the particle lifetime in the trap is of the order

$$\tau \sim \tau_0\,\xi_M^{-2} \approx \frac{\tau_0}{\nu^2}\, e^{2a\epsilon/\nu} \lesssim \frac{\tau_0}{\epsilon^4}\, e^{2a/\epsilon},$$

$$\tau_0^{-1} \sim \Omega\xi^2 \sim \Omega\epsilon_B^{-2}\, e^{-2/\epsilon_B}, \tag{28}$$

where τ_0 is the lifetime at $K > 1$ or $\nu \gtrsim \epsilon$. If one takes account of the dependence $\tau_0(\epsilon_B)$ ($\epsilon_B \sim \epsilon$, $\theta_0 \sim 1$) the lifetime $\tau(\epsilon)$ decreases with $1/\epsilon$ down to the minimal

$$\tau_{\min} \sim \frac{\Omega}{\epsilon^2\nu^2} \exp\left(4\sqrt{\frac{a}{\nu}}\right) \sim \frac{\Omega}{\epsilon^6}\, e^{a'/\epsilon} \quad (a' \sim 1)$$

at maximal $\epsilon \sim \sqrt{\nu}$ (25). The minimal critical value now is still less $K_c(\nu) \sim \epsilon\sqrt{\nu} \gtrsim \epsilon^2 \sim \nu$ (cf. (24)). The last estimate is the condition for a multiplet formation (25).

A considerable decrease in the instability threshold illustrates the general rule: the lower the modulation frequency Ω_M of perturbation, the easier it is for chaos to occur even though it may be confined within a multiplet, for example. The reason is very simple: upon splitting in a multiplet, the width of each resonance $\Delta\omega_r \propto \Omega_M^{\frac{1}{4}}$, and the overlap parameter $\Delta\omega_r/\Omega_M \propto \Omega_M^{-\frac{3}{4}}$ grows indefinitely as $\Omega_M \to 0$.

(c) A weak asymmetry, the Arnol'd diffusion

Is there any loss of particles for $\nu \ll \epsilon^2$ ($K \ll 1$) or for $K \ll K_c(\nu)$? A fine mechanism of the weak instability in many-dimensional nonlinear oscillations has been discovered by Soviet mathematician Arnol'd (1964), Kolmogorov's disciple. As further investigations revealed (Gadiyak et al. 1977; Chirikov 1979; Lichtenberg & Lieberman 1983) Arnol'd's mechanism leads to a slow diffusion along narrow chaotic layers around the destroyed separatrices of nonlinear resonances. This process has been termed *Arnol'd diffusion*. Its most interesting and important feature is *universality*, as chaotic layers exist at any arbitrarily weak perturbation because the frequency of phase oscillations at a resonance $\Omega_r \to 0$ towards the separatrix.

Here again we come across a typical example of '*inverse adiabaticity*' for stationary oscillations driven by high-frequency perturbation. The estimate (27) still holds for the resonant adiabaticity parameter ξ_A too, where now $\Delta\omega = 4\Omega_r$ is the full width of a nonlinear resonance. However, the width can be arbitrarily small as well as ξ_A. Because (see (14) and (15), $\theta_0 \sim 1$)

$$\Omega_r \sim \Omega\sqrt{K} \sim (\Omega/\epsilon_B)\, e^{-1/2\epsilon_B} \tag{29}$$

then for $\nu \ll \epsilon^2$ (no multiplet)

$$\xi_A \sim \nu \exp(f\epsilon_B\, e^{1/2\epsilon_B}); \quad f \sim 1. \tag{30}$$

The first estimate of the destruction ('splitting') of separatrix by a high-frequency perturbation was due to Poincaré (1899). Recently this problem has been studied by many authors (see, for example, Lichtenberg & Lieberman 1983; Zaslavsky 1985), including an accurate evaluation of the full width of a separatrix chaotic layer (Chirikov 1979; Escande 1985).

From (30), a rough estimate for the particle lifetime τ, similar to (28), can be obtained in the form

$$\tau\Omega\nu^2 \sim \exp(2f\epsilon_B\, e^{1/2\epsilon_B}). \tag{31}$$

A distinctive feature of the Arnol'd diffusion in a magnetic trap is the tremendous drop of its rate ($ca.\ \tau^{-1}$) with a slowness parameter ϵ_B of the adiabatic perturbation (a double exponential).

At very small ϵ_B the diffusion is driven by some high-order resonances whose detuning $\delta\omega \ll \Omega$ while their strength is only(!) exponentially small. As a fairly simple analysis shows (Chirikov 1978; for details see Chirikov 1979) the latter mechanism considerably reduces the lifetime

$$\tau\Omega\nu^2 \sim \exp(b\, e^{1/2q\epsilon_B}); \quad q = N, \tag{32}$$

where $N = 3$ is the number of degrees of freedom of the particles and $b \sim 1$ only weakly depends on the parameters.

Arnol'd diffusion has been carefully studied on some simple models (Gadiyak *et al.* 1977; Chirikov 1979; Lichtenberg & Lieberman 1983; Petrosky 1984; Kaneko *et al.* 1985) including the case of high-order driving resonances, or Nekhoroshev's régime (Chirikov *et al.* 1979). A rigorous upper limit for the diffusion rate has been derived by the Soviet mathematician Nekhoroshev (1977), Arnol'd's pupil. The latter estimate can be reduced to the form of (32) with one important difference, namely, for $N \gg 1$ the exponent $q \sim N^2$ rather than $q = N$ in (32). This interesting question remains open. At any rate, in the numerical experiments on a simple model, made by V. V. Vecheslavov as a continuation of the previous work (Chirikov *et al.* 1979), the dependence like (32) was observed down to the extremely weak perturbation which would correspond to $\epsilon_B \approx 0.07$ and $\xi \sim 10^{-5}$.

Arnol'd diffusion was apparently observed also in the laboratory experiments on long-term confinement of electrons in magnetic traps (Ponomarenko *et al.* 1968; Ilyin *et al.* 1977). In any event, the dependence $\tau(B)$ had the characteristic shape of two plateaux with a sharp jump in between. The upper plateau at $B > B_c$ (the critical magnetic field) is determined by the residual gas scattering of an electron on angle $\Delta\theta \sim 1$. The lower plateau is caused by the same scattering but in the nearest chaotic layer of a nonlinear resonance that is on a substantially smaller angle. The latter results in the decrease of electron lifetime by one order of magnitude approximately. Hence, some external noise (scattering) is of great importance for the Arnol'd diffusion because otherwise it would take place for very special initial conditions only, namely, within the chaotic layers whose total relative measure is negligibly small (of the order of (32)). The same is true for the modulational diffusion too, yet the relative layer width here is much bigger ($\sim m\Omega_d/\Omega \sim \nu/\epsilon \gtrsim \epsilon$). One may say also that such a diffusion greatly amplifies the effect of the scattering or of any other noise.

Notice that the critical field $B_c = \omega_c$ is nearly independent of the asymmetry ν (see (12)):

$$\frac{2}{9\pi} \frac{L\omega_c}{\sqrt{\lambda v}} \approx \ln \ln (\tau\Omega\nu^2) - \ln b. \tag{33}$$

Here τ is of the order of the gas scattering time.

Thus, in spite of the universal Arnol'd diffusion the adiabatic invariance proves to be fairly precise generally, and even absolute (eternal) for most initial conditions.

I am deeply grateful to all my colleagues from many countries for a permanent collaboration, in one way or another, which is vital for the progress in this new exciting field of research.

References

Alekseev, V. M. & Yakobson, M. V. 1981 *Physics Rep.* **75**, 287–325.
Arnol'd, V. I. 1963 *Usp. mat. Nauk* **18** (6), 91–192.
Arnol'd, V. I. 1964 *Dokl. Akad. Nauk SSSR* **156**, 9–12.
Bora, D., John, P. I., Saxena, Y. C. & Varma, R. K. 1980 *Plasma Phys.* **22**, 653–662.
Budker, G. I. 1982 *Collected Works*, pp. 72–90. Moskva: Nauka.
Channon, S. R. & Lebowitz, J. L. 1980 *Ann. N.Y. Acad. Sci.* **357**, 108–129.
Chirikov, B. V. 1960 *Plasma Phys.* **1**, 253–260.

Chirikov, B. V. 1978 *Fiz. plasmy* **4**, 512–541.
Chirikov, B. V. 1979 *Physics Rep.* **52**, 263–379.
Chirikov, B. V., Ford, J. & Vivaldi, F. 1979 *A.I.P. Conf. Proc.* **57**, 323–340.
Chirikov, B. V. & Shepelyansky, D. L. 1981 In *Proc. 9th Int. Conf. on Nonlinear Oscillations* (ed. Yu. A. Mitropolsky), vol. 2, pp. 421–424. Kiev: Naukova Dumka.
Chirikov, B. V. 1984 In *Topics in plasma theory* (ed. B. B. Kadomtsev), pp. 3–73. Moskva: Energoatomizdat.
Chirikov, B. V. & Shepelyansky, D. L. 1974 *Physica* D **13**, 395–400.
Chirikov, B. V., Lieberman, M. A., Shepelyansky, D. L. & Vivaldi, F. M. 1985 *Physica* D **14**, 289–304.
Chirikov, B. V. 1986 Asymptotic Methods in Adiabatic Problems. Preprint no. 86–22. Novosibirsk: Institute of Nuclear Physics. (In Russian.)
Chirikov, B. V. & Shepelyansky, D. L. 1986 The Chaos Border and Statistical Anomalies. Preprint no. 86–174. Novosibirsk: Institute of Nuclear Physics. (In Russian.)
Dykhne, A. M. & Chaplik, A. V. 1961 *Zh. eksp. teor. Fiz.* **40**, 666–669.
Escande, D. F. 1985 *Physics Rep.* **121**, 165–261.
Filonenko, N. N., Sagdeev, R. Z. & Zaslavsky, G. M. 1967 *Nucl. Fusion* **7**, 253–266.
Gadiyak, G. V., Izrailev, F. M. & Chirikov, B. V. 1977 In *Proc. 7th Int. Conf. on Nonlinear Oscillations, Berlin, 1975*, vol. II–1, pp. 315–323. Berlin: Akademie-Verlag. (In Russian.)
Galton, F. 1889 *Natural inheritance*. London.
Gelfand, I. M., Graev, M. I., Zueva, N. M., Mikhailova, M. S. & Morozov, A. I. 1963 *Dokl. Akad. Nauk SSSR* **148**, 1286–1289.
Goward, F. K. 1953 In *Lectures on the Theory and Design of an Alternating Gradient Proton Synchrotron*, pp. 19–30. Geneva: CERN.
Greene, J. M. 1979 *J. math. Phys.* **20**, 1183–1201.
Hastie, R. J., Hobbs, G. D. & Taylor, J. B. 1968 In *Proc. 3rd Int. Conf. on Plasma Physics and Controlled Nuclear Fusion Research*, vol. 1, pp. 389–402. Vienna: IAEA.
Hine, M. G. N. 1953 In *Lectures on the Theory and Design of an Alternating Gradient Proton Synchrotron*, pp. 69–82. Geneva: CERN.
Ilyin, V. D. & Ilyina, A. N. 1977 *Zh. eksper. teor. Fiz.* **72**, 983–988.
Kadomtsev, B. B. & Pogutse, O. P. 1979 In *Proc. 7th Int. Conf. on Plasma Physics and Controlled Nuclear Physics Research*, vol. 1, pp. 649–682. Vienna: IAEA.
Kaneko, K. & Bagley, R. J. 1985 *Phys. Lett.* A **110**, 435–440.
Kapitsa, S. P. & Melekhin, V. N. 1969 *The microtron*. Moskva: Nauka.
Karney, C. F. F. 1983 *Physica* D **8**, 360–380.
Karney, C. F. F., Rechester, A. B. & White, R. B. 1982 *Physica* D **4**, 425–438.
Kolomensky, A. A. 1960 *Zh. tekh. Fiz.* **30**, 1347–1354.
Lichtenberg, A. J. & Lieberman, M. A. 1983 *Regular and stochastic motion*. New York: Springer-Verlag.
MacKay, R. S. & Percival, I. C. 1985 *Communs math. Phys.* **98**, 469–504.
Mandelstam, L. I. 1948 *Collected Works*, vol. 1, pp. 297–304. Moskva: Akad. Nauk SSSR.
Nekhoroshev, N. N. 1977 *Usp. mat. Nauk* **32** (6), 5–66.
Petrosky, T. Y. 1984 *Phys. Rev.* A **29**, 2078–2091.
Petrosky, T. Y. 1986 *Phys. Lett.* A **117**, 328–332.
Poincaré, H. 1899 *Les méthodes nouvelles de la mécanique céleste*, vol. 3, §401. Paris.
Ponomarenko, V. G., Trajnin, L. Ya., Yurchenko, V. I. & Yasnetsky, A. N. 1968 *Zh. eksp. teor. Fiz.* **55**, 3–13.
Rechester, A. B. & Rosenbluth, M. N. 1978 *Phys. Rev. Lett.* **40**, 38–41.
Rosenbluth, M. N., Sagdeev, R. Z., Taylor, J. B. & Zaslavsky, G. M. 1966 *Nucl. Fusion* **6**, 297–300.
Vaskov, V. V., Gurevich, A. V., Karfidov, D. M. & Sergeichev, K. F. 1984 *Zh. eksp. teor. Fiz.* (Pisma) **40**, 101–103.
Veksler, V. I. 1944 *Dokl. Akad. Nauk SSSR* **43**, 346–348.
Vivaldi, F. M. 1984 *Rev. mod. Phys.* **56**, 737–754.
Yamaguchi, Y. & Sakai, K. 1986 *Phys. Lett.* A **117**, 387–393.
Zaslavsky, G. M. 1985 *Chaos in Dynamic Systems*. New York: Harwood.

Semi-classical quantization, adiabatic invariants and classical chaos

By W. P. Reinhardt and I. Dana†

*Department of Chemistry and Laboratory for Research on the Structure of Matter,
University of Pennsylvania, Philadelphia, Pennsylvania 19104, U.S.A.*

Einstein–Brillouin–Keller (E.B.K.) quantization of multidimensional non-separable classical hamiltonian dynamics is expeditiously carried out with classical adiabatic switching as hoped for in the early days of the old quantum theory by Ehrenfest, and reintroduced rather recently by Solov'ev. The method is briefly described along with a critique relating to the fact that the method seems to assume that invariant tori exist as a continuous one – parameter function of the coupling constant, although they do not. None the less, the method not only appears to work well for the types of systems envisaged by Ehrenfest, and Solov'ev, but continues to provide a useful asymptotic method following the onset of chaos. After distinguishing between adiabatic and geometric invariants, an analysis of adiabatic switching for the standard map is presented, with results given in the perturbative and strong coupling régimes. In the latter, a key phenomenological result is that principal non-adiabaticities can be understood in terms of the Farey organization of rationals spanning the relevant range of winding numbers. Finally the self-similar geometry of adiabatically generated 'pseudoinvariant' curves is shown from the map of Siegal and Henon.

1. INTRODUCTION

Adiabatic switching is an old idea. In 1911 Einstein is quoted (Jammer 1966) as saying, in response to Lorentz asking about a quantized pendulum retaining its quantization,

If the length of the pendulum is changed infinitely slowly, its energy remains equal to $h\nu$ if it was originally $h\nu$.

Ehrenfest (1916), whose interest in classical adiabatic invariants was stimulated by their use in the derivation of the Wien displacement law, suggested the general classical analogue dubbed 'adiabatenhypothese' by Einstein (1914),

If a system be affected in a reversible adiabatic way, allowed motions are transformed into allowed motions.

The elucidation of these ideas played an important role in the development of the old quantum theory (see Jammer 1966; Klein 1964).

Einstein (1917), without specific reference to adiabatic switching, suggested what has become the modern approach to semi-classical quantization of integrable and nearly integrable non-separable bounded hamiltonian systems. To obtain a

---

† Present address: School of Mathematical Sciences, Queen Mary College, University of London, Mile End Road, London E1 4NS, U.K.

canonically invariant quantization rule he proposed the use of the Poincaré invariant form $\sum_i p_i \, dq_i = \boldsymbol{p} \cdot d\boldsymbol{q}$, with the N independent quantization conditions (for a systems of N freedoms)

$$\frac{1}{2\pi} \oint_{C_i} \boldsymbol{p} \cdot d\boldsymbol{q} = (n_i + \tfrac{1}{4}\alpha_i) \, \hbar \tag{1}$$

being calculated on N topologically independent one-cycles, C_i, on the N-tori assumed to exist everywhere in the phase space. The $\frac{1}{4}\alpha$ is a correction for caustics introduced by Brillouin (1924) and Keller (1958), giving rise to the acronym E.B.K. for such quantization on invariant tori (see Percival 1977). It is important to note that the paths C_i are not classical trajectories, except in the special case $N = 1$ where the tori, and integration paths have both degenerated to become identical with single periodic oribts, whereupon the quantization conditions reduces to

$$\frac{1}{2\pi} \oint p \, dq = T/\pi = (n + \tfrac{1}{4}\alpha) \, \hbar, \tag{2}$$

where T is the kinetic energy integrated over one period of the classical motion, that is $\int (p^2(t)/2m) \, dt$, and is precisely Ehrenfest's primary adiabatic invariant, and thus perhaps the simplest example of the connection between quantization and adiabatic invariance.

In what follows we briefly discuss the differing geometric content of adiabatic invariance for systems of one and N freedoms. In particular, in §2, the distinction between adiabatic and geometric invariant actions is made. In §3 adiabatic semiclassical quantization on tori is described, and shown to provide reasonable estimates of eigenvalues even in the presence of classical chaos, where the tori envisaged by Einstein and Solov'ev have surely ceased to exist. The question, raised earlier by Berry (1984), of why the method can work at all, even when tori exist in large measure, is addressed in §4, where specific criteria for adiabatic approximation of tori are found in an analysis of the standard map. Section 5 raises the issue of the geometric interpretation of the success of the method for quantization of chaos, by introduction of the new concept of pseudoinvariants, which are found to have phase space structures on all scales. The paper concludes with a discussion.

2. ADIABATIC AND GEOMETRIC INVARIANCE

Systems of one-freedom

The adiabatic invariance of T, for one-freedom systems, is usually given the pleasing geometric interpretation of the area (see, for example, Born 1966)

$$2T = \oint p \, dq \tag{3}$$

being conserved under adiabatic changes in the hamiltonian generating the time evolution of p and q. Namely it is assumed that

$$\dot{p} = -(\partial H/\partial q)(q, p, a), \tag{4a}$$

$$\dot{q} = (\partial H/\partial p)(q, p, a), \tag{4b}$$

with $a = a(t)$ being a slowly varying parameter in the definition of H. For the interpretation of (3) to be true at any 'instant' of time it must be the case that the period of motion, over which the area, equal to $2T$, is calculated is fast compared with the rate of change of $a(t)$ so that over one period $a = a(t)$ hardly changes. Corrections to this geometric picture are discussed by Percival & Richards (1982). In other words, the success of adiabatic methods relies on the fact that averaging (see Arnol'd 1978a) takes place over the whole of the allowed phase space (the energy shell), which is one dimensional (because of energy conservation on the time scale of one period) for one-freedom systems, and thus fully explored by the trajectory during each period of motion.

Systems of N-freedoms

Many aspects of the above description change profoundly for systems of more than one freedom. The most important differences for the present discussion are that: a single trajectory no longer always fills the energy shell; and the paths, C_i, in the quantum conditions of (1) are not, as mentioned earlier, time evolving trajectories, but rather time-independent curves on the invariant manifold of trajectories defined by the torus. The second of these differences forces upon us a distinction, not usually made, which requires us to distinguish carefully between preservation of the 'area', or

$$\oint_\gamma \boldsymbol{p} \cdot \mathrm{d}\boldsymbol{q},$$

and non-adiabaticity. This distinction arises in the numerical implementation of the E.B.K. quantization of multidimensional systems, and thus we face it at once.

Suppose we take a path, γ, defined as a one-cycle on an invariant torus of the hamiltonian $H(\boldsymbol{q}, \boldsymbol{p}, a(t = 0))$. Such a closed path may be thought of as defining a one parameter family of initial conditions $\{\boldsymbol{q}(\gamma), \boldsymbol{p}(\gamma)\}$ at $t = 0$. These initial conditions, and specification of the hamiltonian $H(\boldsymbol{q}, \boldsymbol{p}, a(t))$ uniquely define the image of γ at time $t = T$, which we call γ_T. It is now a geometric fact, equivalent to Stokes's theorem (Arnol'd 1978b), that

$$\oint_\gamma \boldsymbol{p} \cdot \mathrm{d}\boldsymbol{q} = \oint_{\gamma_T} \boldsymbol{p} \cdot \mathrm{d}\boldsymbol{q} \tag{5}$$

for any time-dependent hamiltonian whatsoever. Thus the E.B.K. quantization conditions, as imposed by direct (pointwise) time evolution of the integration path, γ, are geometric invariants, whether or not the switching is adiabatic, or whether or not the system is even integrable. The relevance of this simple remark, illustrated in figure 1, to the way in which actual calculations are carried out will become clear in the next section. Adiabaticity or non-adiabaticity then relates to whether the new path, γ_T, lies on an invariant manifold of the hamiltonian $H(q, p, a(t = T))$, not to whether the phase-space line integral (or area) is preserved, as it always is!

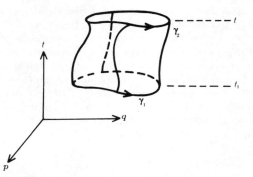

FIGURE 1. Evolution of a manifold of trajectories, defined at time t_1, by the curve of initial conditions, γ_1. In the extended (q, p, t) phase space, the curve evolves into its image, γ_2, at time t_2, via evolution of the trajectories, each of which is governed by the same time-dependent hamiltonian. Then, for a general time-dependent hamiltonian (i.e. not necessarily adiabatic!),

$$\oint_{\gamma_1} p \cdot dq = \oint_{\gamma_2} p \cdot dq.$$

The importance of this remark is that the use of time-dependent switching to generate a final manifold by propagation of a swarm of classical trajectories guarantees satisfaction of the E.B.K. quantization conditions, whether or not the final manifold is an invariant of the final hamiltonian. If the adiabatic hypothesis holds the final manifold *is* such an invariant surface. The adiabatic methods discussed here thus involve averaging in the extended phase space to the extent that adiabaticity fails, but the E.B.K. conditions are satisfied exactly on the approximate tori.

3. Adiabatic implementation of E.B.K. quantization

Solov'ev (1978), and independently Johnson (1983), have suggested that an expeditious manner to carry out the E.B.K. process for quantization on tori would be to pick a separable (or analytically integrable) unperturbed hamiltonian, H^0, and adiabatically switch on a perturbing term, $V(q)$, via a switching function $a(t)$, such that $a(0) = 0$, *and* $a(T) = 1$. Thus

$$H(q, p, a(0)) = H^0(q, p), \qquad t = 0, \qquad (6a)$$

$$H(q, p, a(t)) = H(q, p) + a(t) V(q), \quad 0 \leqslant t \leqslant T, \qquad (6b)$$

$$H(q, p, a = 1) = H(q, p) + V(q), \qquad t > T. \qquad (6c)$$

If the invariant tori of H^0 are easily found, those which satisfy the E.B.K. quantization conditions may be used to define $t = 0$ initial conditions of trajectories, which are assumed to lead for times $t > T$ to motion on the invariant tori of H, itself. Further as the actions of (1) are preserved under evolution via the explicity time dependent hamiltonian, $H(q, p, a(t))$, via (5), the E.B.K. conditions will be satisfied.

The Ehrenfest adiabatic assumption is then that the initial invariant manifold of H^0 is converted into an invariant manifold of H. If this is so the resulting trajectory will lie on precisely that invariant torus that satisfies the E.B.K. quantization conditions, and thus the semi-classical eigenvalue will be given by $E(T) = H(q, p, a = 1)$. A visualization of this process is shown in figure 2, where an initial invariant torus of H^0 is shown on the left, with the 'corresponding' torus of H shown on the right. Can we visualize adiabatic switching as the smooth distortion of the initial torus into the final torus, as a function of the monatonic switching function $a(t)$? If by this we assume that at each value of the coupling constant, $a(t)$, along the switching process, that trajectories begun on the tori of H^0 will define a new torus, intermediate between those of figure 2, we are in error.

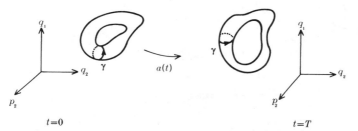

$t=0$ $t=T$

FIGURE 2. Visualization of adiabatic quantization on tori. At $t = 0$ an E.B.K. torus associated with the dynamics of the *unperturbed* dynamics is easily found. For times between 0 and T a coupling, parametrized by $a(t)$, is turned on, and the torus deforms, ending, at $t = T$, as an invariant manifold of the full dynamics, but with the E.B.K. conditions still satisfied. This simple picture cannot correspond to reality, as discussed in the text and illustrated in figure 3, as there is, in general, no continuous family of tori lying between the indicated initial and final tori desired in the adiabatic process.

This is simply because as the coupling changes, the ratios of frequencies of evolution of the angle variables, $\theta_i = \omega_i t + \delta_i$, will pass through rational values, and the resulting periodic or resonance orbits will not define tori. In fact, the situation is still worse: associated with the existence of these periodic orbits there will, in general, exist resonance zones and associated bands of chaos. Berry (1984) has suggested that the method can only succeed to the extent that the switching is fast enough to pass rapidly through the (dense) sets of rational winding number ratios. We present an analysis of this phenomenon in §4. However, as established by Solov'ev (1978), Johnson (1983, 1985), Hedges *et al.* (1984), Skodje *et al.* (1985), Grozdanov *et al.* (1986) and Patterson (1985), the method works well for quantization where such resonance zones, and associated chaotic regions are of small measure, and in fact Skodje *et al.* (1985) have shown that the method works, at least asymptotically, past the destruction of tori, that is in the regions of largescale chaos. This is shown in figure 3.

Before giving an analysis of the success these applications, and before attempting to quantify its range of applicability, it should be noted that all of the above workers carry out the quantization by defining a family of initial conditions on the appropriate tori of H^0, propagating the entire family of trajectories forward in time according to the hamiltonian of (6), followed by averaging $E(t)$ over this

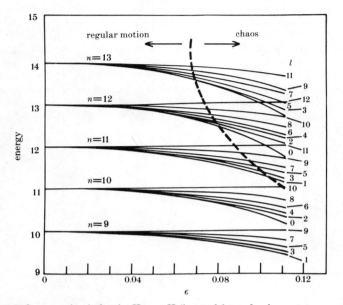

FIGURE 3. Quantum levels for the Henon–Heiles model two-freedoms system as a function of coupling constant, as obtained by adiabatic switching as discussed in §3. The dashed line indicates quantum levels in the regions of phase space where the classical dynamics is chaotic, and thus the final E.B.K. tori cannot exist. Yet the method seems to continue to work well, as the displayed levels are in excellent agreement with full quantum results. Skodje *et al.* (1985).

ensemble during the switching process. In a sense this approach attempts to map the initial manifold into the final on a point by point basis: if the quantization conditions of (1) are tested on this approximation to the final manifold, they are found to be exactly satisfied, as in the discussion of §2, whether or not the manifold obtained is an invariant of the final dynamics. This is clearly a strength of the method.

4. ADIABATIC SWITCHING AND THE STANDARD MAP

Generalities

To study the applicability of adiabatic switching methods to non-integrable hamiltonian systems, which will be the generic case for systems with more than one degree of freedom, even in those regions where invariant tori exist in large measure, we resort to the use of area preserving maps. These algebraic recursions mimic the behaviour of the Poincaré sections of two-freedom hamiltonian systems, and thus display many of the features which differentiate one-freedom and *N*-freedom systems. A useful such map, originally due to Chirikov, has been studied by Green (1979), Shenker & Kadanoff (1982), and by MacKay *et al.* (1984, 1986), and has been dubbed the 'standard map' by Percival. We thus consider the recursion, where I and θ are thought of as action-angle variables:

$$I_{n+1} = I_n + K \sin (\theta_n), \qquad\qquad (7a)$$

$$\theta_{n+1} = \theta_n + I_{n+1}. \qquad\qquad (7b)$$

Figure 4 gives an idea of the complexity of the phase space for an intermediate value of the perturbation K. There appear to be some single valued and continuous curves, $I(\theta)$, running from $\theta = 0$ to $0 = 2\pi$. These are K.A.M. (Kolmogorov–Arnol'd–Moser) surfaces, or 'rotational' tori, which may be taken to be functions of θ. Other regions of the phase space are characterized by resonance islands, whose centres are periodic orbits, of increasing length. Surrounding the resonance islands are bands of separatrix chaos. At $K = 0$ the analogous picture reduces to a series of straight lines, $I(\theta) = $ constant, where the constant is given by 2π times the winding number

$$\omega(I) \equiv \lim_{n \to \infty} \left(\frac{\theta_n - \theta_0}{2\pi n} \right). \qquad\qquad (8)$$

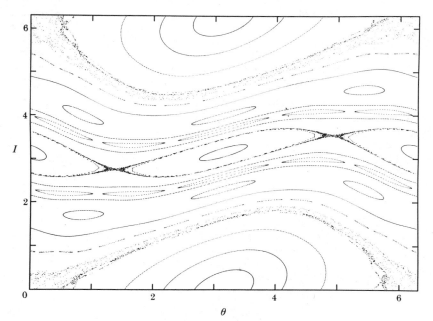

FIGURE 4. Phase-space structure for the standard map at $K = 0.80$. Resonance zones and island chains are clearly seen, in addition to bands of chaos. At smaller K, at appropriately higher magnification, the general structure of phase space would be similar for this non-integrable system. As island chains are associated with *every* rational winding number, we can ask why the adiabatic switching method ever works, as the desired tori certainly do not exist along the switching process, even if the final E.B.K. torus *does* exist.

For smaller but non-zero K, rotational tori predominate, but on a renormalized scale, the phase space would look about like that of figure 4, with islands and separatrix chaos interspersing the rotational tori, each rational winding number $\omega = p/q$ generating an island chain.

The problem of adiabatic switching for the standard map, which we imagine is representative of a generic two-freedoms system, is now easily seen. During the course of the slow switching process, it becomes evident that the winding number is a function of the coupling constant at constant action. As the winding number changes, it passes infinitely often through rational values, and thus, at these values of K, the rotational tori no longer exist, being replaced by resonance tori, with their bands of surrounding separatrix chaos. It is thus not surprising that detailed numerical experiments (Dana & Reinhardt 1987 a, b) show that even for very small final values of K, that adiabatic switching leads to multivalued functions of θ, exhibiting 'whorls' and 'tendrils' earlier seen by Berry et al. (1979).

In the next subsection we outline a perturbative analysis of the switching process, valid for small K, and which allows us to see the role of the choice of switching function on the rate of convergence of the result. This is followed by a qualitative discussion of switching to larger K, where the method becomes of necessity asymptotic. Results of an empirical analysis of Dana & Reinhardt (1987 a, b) allowing prediction of which rational periodic orbits will result in the dominant contribution to intrinsic non-adiabatic behaviour, and predictions of optimal switching times for approximation of rotational tori are summarized.

Perturbative analysis of adiabatic switching

Results valid for small K may be obtained perturbatively on the assumption that the adiabatic switching process actually leads to a rotational torus. For the standard map, all rotational tori are single-valued functions of θ, and thus the adiabatically switched standard map,

$$I_{n+1} = I_n + Ks(n) \sin (\theta_n), \tag{9a}$$
$$\theta_{n+1} = \theta_n + I_{n+1}, \tag{9b}$$

can be solved for $I(\theta)$ order by order in K by iteration, given a specific choice of switching function $s(n)$, which is the analogue of the continuous function $a(t)$ used in the discussion of continuous time hamiltonian dynamics. N (the analogue of T) gives the number of iterations of the map, during which the coupling strength is increased from 0 to K. For example, taking

$$S(n) = \frac{n}{N} - \frac{\sin (2\pi n/N)}{2\pi}, \tag{10}$$

which is the discrete time analogue of the switching function used in many of the references in §3, we find

$$I(\theta) = I^0 - \frac{K \cos (\theta - \tfrac{1}{2}I^0)}{2 \sin (\tfrac{1}{2}I^0)} \tag{11a}$$

with an RMS error in the action given by

$$\frac{\pi^2 K}{\sqrt{(2\pi)}16 \sin^4 (\tfrac{1}{2}I^0)} \frac{|\sin (\tfrac{1}{2}NI^0)|}{N^3} + O(N^{-5}). \tag{11b}$$

The N independent terms are the zeroth and first perturbative approximations to the fixed action rotational torus of action I^0, and the first-order approximation to the effect of generating the torus via adiabatic switching suggests that such effects vanish as N^{-3}. As the switching function $s(n)$ has two continuous derivatives for n near 0 and also for n near N, the final switching 'time', this is not an accident. Application of the Euler–Maclauren sum rule gives the general result that if $s(n)$ has M continuous derivatives near the endpoints of the switching process, then we can expect an $N^{-(M+1)}$ dependence of non-adiabatic contributions to the desired rotational torus. If exponential convergence is desired, $s(n)$ needs to have essential singularities such as $\exp(-t/n)$ near $n = 0$, or $\exp(-1/(N-n))$ near $n = N$. In practice, use of such singular switching functions does give the expected exponential convergence for large values of N, but for small values of more practical interest actually gives results which converge more slowly than for 'less optimal' switching functions giving *only* inverse-power-law convergence. Such a perturbative analysis fails for large N, in any case, as is well illustrated in the following subsection.

Intrinsic non-adiabaticity

Figure 5a, b shows the result of switching to the golden-mean torus (i.e. that torus with the 'least rational' winding number, in the sense of rational approximation, which is the golden frequency $\tfrac{1}{2}(\sqrt{5}-1)$) at the critical value of $K = 0.971\,635\ldots$, beyond which rotational tori no longer exist (see Greene 1979). In (a), the switching time $N = 233$ has been used, resulting in an excellent approximation to the actual golden-mean torus. In (b), $N = 1597$, which, inspite of the fact that the switching is slower gives a much worse result, in fact one that is not even single valued! The switching thus provides an optimal approximation to a given rational torus for a *finite* value of N, and in this sense is asymptotic. How is this value of N determined, and what rational (resonance) winding number will dominate? Without going into full detail, for which see Dana & Reinhardt (1987a, b), which in turn relies on MacKay *et al.* (1986), the key to such an analysis is to determined $\omega(K, I)$, the winding number as a function of coupling constant, K, and action I. This may be done accurately and conveniently by adiabatic switching! Once this function is known, approximations to a torus of action I may be obtained by finding neighbouring rational pairs p/q and p'/q' (these are neighbours *iff* $pq' - p'q = 1$) which just bracket the full range of variation of $\omega(K)$ for the value of I of interest. The Farey tree (see, for example Hardy & Wright 1954; Kim & Ostlund 1986) constructed in terms of successive Farey means, $(p+p')/(q+q')$, from this initial pair provides a systematic enumeration of all rational in the specified interval, as a function of increasing denominator q, which gives the order of the island chain encountered at a rational value of $\omega(s(n)K)$. Given the expectation that period orbits of very high order (large q) will have smaller effect than those associated with lower-order rationals, this Farey decomposition of the appro-

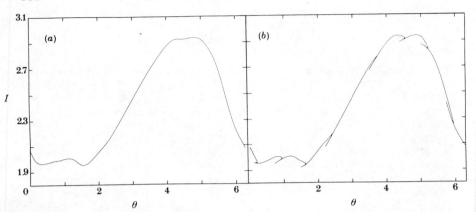

FIGURE 5. (a) Adiabatic approximation to the golden torus (Greene 1979) obtained via adiabatic switching during a period of 233 time steps. This curve obtained at the critical value, $K = 0.97163\ldots$, is almost identical to the exact critical torus, dividing the phase space into two disjoint chaotic regions. Adiabatic switching thus seems to work at the brink of destruction. (b) Approximation to the golden-mean torus at $K = 0.97163\ldots$ obtained by adiabatic switching with 1597 time steps. This slower switching has led to a result of considerably less value than that shown in figure 5a, indicating the asymptotic nature of the method for non-integrable systems. The prediction of the nature of the breakdown of the method, and the choice of optimal switching times is discussed in §4. From Dana & Reinhardt (1987a).

priate segment of the real line, provides a systematic procedure for determining the critical switching time N_{crit} beyond which the adiabatically generated rotational torus will become multivalued. This critical value is determined as the minimum of the $\tilde{N}_q$ for all qs appearing in the Farey approximants to the variation of $\omega(s(n)K)$, where N_q measures the influence of the island chain of order q on the switching. To leading order

$$\tilde{N}_q \sim K^{-1}(s'(n))K^{-q}. \tag{12}$$

The result expressed in (12) has a simple physical interpretation: the derivative $s'(n)$, evaluated for that value of n lending to $\omega(Ks(n)) = p/q$, measures the length of time that $\omega(K)$ remains 'near' the offending rational value, and is dependent on the form of $s(n)$, and in particular shows that resonances will be more important near the end of the switching, where $s(n)$ is varying most slowly. The factor K^{-q} arises as

$$K^{-q} \approx \text{(diffusion time for } p/q \text{ island)/(width of } p/q \text{ island chain)} \tag{13}$$

as follows from the residue analysis of MacKay et al. (1986). We note that for small K and high order rational winding numbers (large q) that the critical switching times will be so long as to completely validate the use of perturbative estimates of the previous subsection for all practical purposes. As long as these critical switching times are compared with the reciprocal level spacings there will be no serious problem for quantization, and this low-level non-integrability is well controlled by the method, as intuitively envisaged by Solov'ev (1978).

In summary, the use of adiabatic switching to obtain asymptotic approximations to rotational tori, with zeroth-order rotational tori as the initial approximation is now fairly well in hand. The use of adiabatic methods to obtain the non-rotational (island chain) tori of figure 4, is a different matter altogether: this more subtle problem has received preliminary discussion by Skodje *et al.* (1985), Grozdanov *et al.* (1986), and much further developed by Martens & Ezra (1986) and Ezra *et al.* (1987) using a Lie-transform-based methodology developed by Farrelly (1986).

5. Chaos

The success of adiabatic switching in quantization of the levels for the Henon–Heiles model problem, figure 3, even into the chaotic regions of phases space, requires that the analysis of §4 be extended to include and understanding of the fate of invariant manifolds adiabatically switched to couplings leading to chaos. A first, empirical, look at this is shown in figure 6 where iterates of an adiabatically obtained torus for the Siegal–Henon map

$$p_{n+1} = p_n \cos \alpha - (q_n - s(n) \, p_n^2) \sin \alpha, \qquad (14a)$$

$$q_{n+1} = p_n \sin \alpha - (q_n - s(n) \, p_n^2) \cos \alpha, \qquad (14b)$$

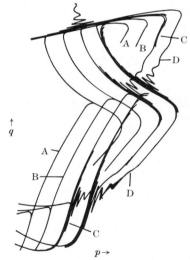

Figure 6. Pseudoinvariant curves obtained by switching on the quadratic nonlinearity of the Henon–Siegal map (14). Curve A, 600 time steps. One further iteration of the curve thus obtained is labelled B, 600 additional iterations resulted in curve C and 1400 additional iterations resulted in curve D. The curves have been successively displaced, to allow visualization of the development of the homoclinic structure. Were such displacement not made, the curves would almost exactly superimpose in successive iterations, giving rise to the sobriquet of *pseudoinvariant*. It seems that if invariants do not exist at the final value of the coupling, adiabatic switching still attempts to find the next best thing. It may be noted that curve D shows overlapping homoclinic oscillations, better seen in figure 7.

are shown, for 600, 601, 1200 and 2000 iterations of the map, where adiabatic switching was carried out during the first 600 iterations, whereupon $s(n)$ was set equal to unity. What is obtained is a *pseudoinvariant*, which retains its general shape and features from iteration to iteration, eventually filling an area, but in a highly regular manner. This regularity is better displayed in figure (7) where, in a polar representation, the angle variable, $\theta = a \tan(p/q)$, is displayed as a function of arclength along the final pseudoinvariant curve obtained at 2000 iterations of the original map. The regularity of the structure is striking, clearly suggesting application of renormalization techniques. Such an analysis is now in progress.

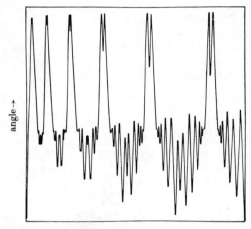

arc length on final pseudoinvariant

FIGURE 7. Pseudoinvariant curve D (of figure 6) displayed as final angle plotted as a function of final arclength. An approximately self-similar structure is evident, suggesting the utility of a scaling analysis. As the number of iterations increases, complexity increases at *all levels* not only on finer and finer scales. Thus the usual fractal dimension saturates at 2, in an uninteresting manner, suggesting a need for a more subtle analysis.

6. DISCUSSION

Adiabatic swtiching provides a powerful method for semi-classical quantization of multidimensional systems. Although at first glance the foundations of the method seem simple, as an unperturbed reference torus is continuously switched to a final quantized torus, further reflection reveals that the method can only be asymptotic, as the envisaged continuous family of tori does not exist: the winding number(s) ratios will be rational on a dense set of real values of the coupling constant, and the intermediate tori then cease to exist, even when the initial and final E.B.K. tori do exist. The nature of this asymptotic method is beginning to be understood.

What about the case when the final torus does not exist, being replaced by chaos on the large scale ? The pseudoinvariant behaviour shown in §5 indicates that there will be cases where a detailed analysis will be possible. As tendrils spread globally into the whole phase space, the energy dispersion, over the adiabatically obtained (non-invariant) manifold can become arbitrarily large, but the timescales over which this happens should set (useful?) bounds on the utility of the method. However, it will also be so that the local energy on this non-invariant (or sometimes pseudoinvariant) manifold will be a relatively slowly varying function of local coordinate, and thus may well contain additional information, beyond the average energy now used in adiabatic semi-classical quantization.

We are grateful for the opportunity to present this material at the Royal Society Discussion Meeting, and thank M. V. Berry, I. C. Percival and N. O. Weiss, and the staff of the Society for their organization of a stimulating meeting. We also acknowledge a prepublication copy of the paper MacKay *et al.* (1986). This work was supported by grants CHE84-16459, and DMR85-19059 from the National Science Foundation.

REFERENCES

Arnol'd, V. I. 1978*a* *Mathematical methods of classical mechanics*, pp. 292–300. New York: Springer.
Arnol'd, V. I. 1978*b* *Mathematical methods of classical mechanics*, pp. 233–240. New York: Springer.
Berry, M. V. 1984 *J. Phys.* A **17**, 1225–1233.
Berry, M. V., Balazs, N. L., Tabor, M. & Voros, A. 1979 *Ann. Phys.* **122**, 26–63.
Born, M. 1966 *Atomic physics*, 7th edn, p. 118. New York: Hafner.
Brillouin, L. 1926 *J. Phys., Paris* **7**, 353–368.
Dana, I. & Reinhardt, W. P. 1987*a* *Physica* D. (In the press.)
Dana, I. & Reinhardt, W. P. 1987*b* In *Proc. 1st International Conference on the Physics of Phase Space (Lecture notes in physics)*. New York: Springer. (In the press.)
Ehrenfest, P. 1917 *Phil. Mag.* **33**, 500–513.
Einstein, A. 1914 *Verh. dt. phys. Ges.* **16**, 820–828.
Einstein, A. 1914 *Verh. dt. phys. Ges.* **19**, 82–89.
Ezra, G. S., Martens, C. C. & Fried, L. E. 1987 *J. phys. Chem.* (In the press.)
Farrelly, D. 1986 *J. chem. Phys.* **85**, 2119–2131.
Greene, J. M. 1979 *J. math. Phys.* **20**, 1183–1201.
Grozdanov, T. P., Saini, S. & Taylor, H. S. 1986 *Phys. Rev.* A **33**, 55–67.
Hardy, G. H. & Wright, E. M. 1954 *An introduction to the theory of numbers*, 4th edn. Oxford: Clarendon.
Hedges, R. M., Skodje, R. T., Borondo, F. & Reinhardt, W. P. 1984 *Am. Chem. Soc. Symp. Ser.* **263**, 323–336.
Jammer, M. 1966 *The conceptual development of quantum mechanics*, ch. 3, sect. 1, pp. 89–108. New York: McGraw-Hill.
Johnson, B. R. 1983 *Aerospace Report*.
Johnson, B. R. 1985 *J. chem. Phys.* **83**, 1204–1217.
Keller, J. B. 1958 *Ann. Phys.* **4**, 180–188.
Klein, M. 1964 In *Proc. 10th Internal. Cong. of History of Science, Cornell, 1962*, pp. 801–804. Paris: Hermann.
Kim, S. & Ostlund, S. 1986 *Phys. Rev.* A **34**, 3426–3434.
MacKay, R. S., Meiss, J. D. & Percival, I. C. 1984 *Physica* D **13**, 55–81.
MacKay, R. S., Meiss, J. D. & Percival, I. C. 1986 Preprint no. 237. Institute for Fusion Studies, University of Texas at Austin.
Martens, C. C. & Ezra, G. S. 1987 *J. chem. Phys.* **86**, 279.
Patterson, C. W. 1985 *J. chem. Phys.* **83**, 4618–4632.
Percival, I. C. 1977 *Adv. chem. Phys.* **36**, 1–61.

Percival, I. C. & Richards, D. 1982 *Introduction to dynamics*, ch. 9. Cambridge University Press.
Shenker, S. J. & Kadanoff, L. P. 1982 *J. statist. Phys.* **27**, 631–656.
Skodje, R. T., Borondo, F. & Reinhardt, W. P. 1985 *J. chem. Phys.* **82**, 4611–4632.
Solov'ev, E. A. 1978 *Soviet Phys. JETP* **48**, 635–639.

Some geometrical models of chaotic dynamics

By C. Series

Mathematics Institute, University of Warwick, Coventry CV4 7AL, U.K.

The free motion of a particle on a surface of constant negative curvature (a pseudosphere) was one of the first models of chaotic motion. It became the prototype for the theory of hyperbolic systems developed by Bowen and Sinai. In these models, geometry suggests a symbolic coding which already exhibits fully chaotic behaviour.

One can return to these models to seek possible manifestations of quantum chaos. Here the mathematical technique is harmonic analysis on hyperbolic space. Chaotic behaviour seems to appear both in the behaviour of individual eigenfunctions and in the sequence of spectral values.

Introduction

The free motion of a particle on a surface of constant negative curvature was probably the first example of what is now described as chaotic motion. Hadamard (1898) made a detailed study of geodesics on such surfaces, and in particular noted the occurrence on certain surfaces of families of geodesics whose cross section exhibits a Cantor- or fractal-like structure. Geodesics on these surfaces diverge exponentially at a constant rate, thus among those geodesics that remain in a bounded portion of the surface (the non-wandering set) one has all the ingredients for fully chaotic motion. Moreover, one has to hand a highly developed mathematical description of these surfaces; in particular their metrical structure is locally that of hyperbolic or non-euclidean geometry.

The idea of coding trajectories on these surfaces by symbol sequences first appears in (Morse 1921, 1966) and (Koebe 1917†, 1929). As we shall explain below, Morse restricted himself to a particularly simple class of examples, whereas Koebe treated fully the most general case. Whereas Morse was motivated by the dynamical analogy, Koebe's work formed part of a systematic study of these surfaces. Both authors used their symbolisms to exhibit trajectories everywhere dense on the surface; later Artin (1924) did the same for the modular surface using a coding of a rather different kind based on continued fractions, and Hedlund (1934, 1935) extended the method to prove ergodicity of the geodesic flow in certain special cases. Much later, these symbolic codings were replaced by the very general methods developed by Sinai (1968), Ratner (1973) and Bowen (1973) for hyperbolic systems.

† The manuscript to which the earlier date refers was a preliminary version of Koebe's prize essay (1917). A footnote in *Acta. Math.* **50** (1927) leads me to believe that this manuscript contains the earliest version of symbolic coding on surfaces of constant negative curvature; unfortunately, the manuscript appears to be lost among the archives of the Mittag Leffler Institute in Stockholm.

THE MODEL

The surfaces in question have the property of being everywhere saddle shaped. They are homogeneous and locally isotropic; in other words, the local geometry at a point is independent both of the point and of the direction chosen. A complete surface of this type cannot be embedded in three-dimensional euclidean space

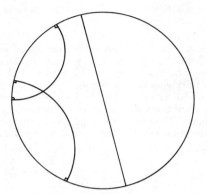

FIGURE 1. The Poincaré disc.

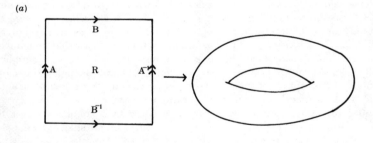

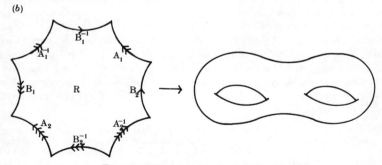

FIGURE 2. Glueing up to form surfaces.

$\mathbb{R}^3$; thus for example the surfaces shown in figure 4 are distorted so that the curvature appears to be positive at some points.

The geometry is best exhibited by projecting onto the euclidean plane in a manner exactly analogous to the more familiar projections of the globe used by cartographers. The projection that I shall describe is known as the *Poincaré disc*. This exhibits simply connected hyperbolic space (analogous to the euclidean plane, and the universal cover of all the surfaces in question), as the interior of the unit disc $\mathbb{D}$ in the complex plane, see figure 1. Geodesic lines appear as circular arcs orthogonal to the boundary circle B; angles are measured in the ordinary euclidean sense and distance is given by $ds = 2|dz|/(1-|z|^2)$. (The curvature here is normalized to be -1.) The isometries (distance-preserving maps) of the plane are given by linear fractional transformations of the form $z \rightarrow (az+b)/(\bar{b}z+\bar{a})$; here $a, b \in \mathbb{C}$ and $|a|^2-|b|^2 = 1$.

The surfaces in question are associated with *tesselations* of hyperbolic space, just as the familiar flat torus is associated to the tesselation of the euclidean plane by unit squares. A tesselation is a covering of the plane by congruent (isometric) copies of some basic figure, the *fundamental region*, with the property that distinct regions intersect only along common boundaries. A surface is formed when those isometries that carry one side of the fundamental region onto a side of an adjacent

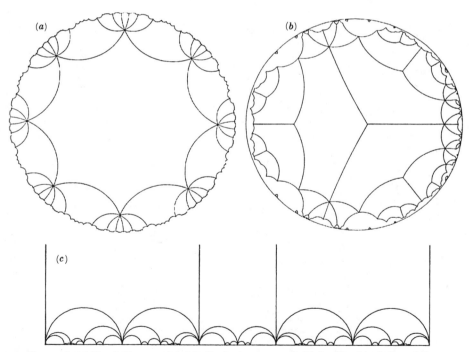

FIGURE 3. (*a*) Tesselation associated to the 'octagon group'. (*b*) Another tesselation of the disc. (*c*) Tesselation of the hyperbolic plane corresponding to the 'leaky torus'.

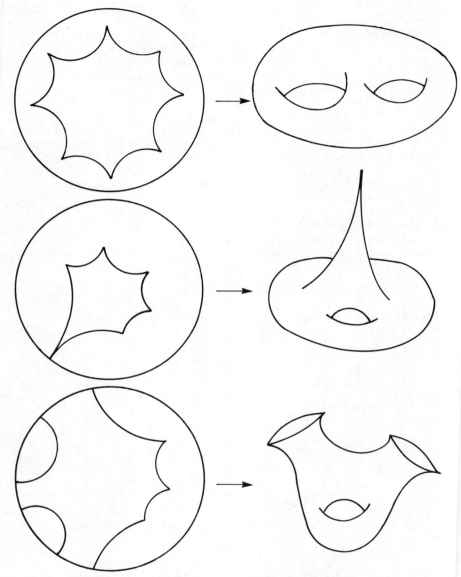

FIGURE 4. The three types of hyperbolic surface: type 1, compact, finite area; type 2, non-compact, finite area; type 3, non-compact, infinite area.

region are used to 'glue' the two sides together (figure 2*b*), just as a torus may be formed by glueing horizontal and vertical sides of a unit square (figure 2*a*). Because the glueing map is an isometry, the local metrical structure of hyperbolic space is preserved.

A great wealth of possible tesselations occurs in hyperbolic space, of which a few are shown in figure 3. (Figure 3*c* is a tesselation of the *upper half plane* model of hyperbolic space, $\mathbb{H} = \{z \in \mathbb{C} : \operatorname{Im} z > 0\}$.) Indeed, any surface with at least two holes (genus $\geqslant 2$) and any number of boundary curves and spikes, may be realized in this way. Among surfaces formed from finite-sided fundamental regions, three main types may be distinguished, shown in figure 4. Surfaces of the second type are non-compact but have finite area; geodesic motion on these surfaces is still ergodic. Surfaces of the third type, with infinite funnels, are those on which Hadamard observed his Cantor-like sets of geodesics; these were geodesics that remain forever in a bounded portion of the surface without escaping to infinity down the funnel. These are also the surfaces to which Morse's analysis (Morse 1921) applies. (In addition, Morse allowed variable negative curvature.) Koebe had a coding that even applied to certain fundamental regions with an infinite number of sides.

THE CODING

The basic idea of the coding that I describe is very simple, and is a modification of the methods of Morse and Koebe. Take a particular surface with associated fundamental region R (figures 2*b* and 3*a*). The sides of R are geodesic arcs.

Each side of R is identified with another such by an isometry of $\mathbb{D}$; in the particular case illustrated we have four such isometries A_1, B_1, A_2, B_2 together with their inverses. The sides of R are labelled by these eight elements Γ_R, as shown.

Now a geodesic on the surface can be drawn on R in the following way. We follow a geodesic arc across R until we meet some side; this side is identified to another side by some $X \in \Gamma_R$, and so the geodesic in question reappears entering R across the side labelled X^{-1}. Continuing in this way we obtain a collection of arcs across R, as shown in figure 5, which is finite if and only if the corresponding geodesic is closed. The coding is obtained simply by listing, in order, the labels of the sides traversed. Thus we associate to each geodesic a bi-infinite sequence of symbols in Γ_R, which is periodic if and only if the geodesic is closed. An alternative description is that we have taken unit tangent vectors based on the sides of R as a surface of section to the motion (regarded as motion on the unit tangent bundle of the surface.)

We must now ask which sequences of symbols in Γ_R arise in this way. We note immediately that blocks XX^{-1} are excluded, for this would indicate a geodesic entering and immediately leaving across the same side of R, which is impossible from the geometry of geodesic lines in $\mathbb{D}$. In general, certain other exclusions arise, associated to the vertices of R in $\mathbb{D}$. However, if all vertices of R lie on B; we obtain (Series 1986) the following.

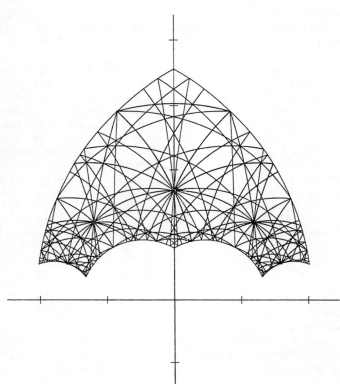

FIGURE 5. A geodesic on the surface corresponding to the tesselation in figure 3*b*.

THEOREM *A sequence occurs as a trajectory of some geodesic on the surface, if and only if no block* XX^{-1} *occurs.*

In general, the exclusions related to the vertices arise for reasons such as those illustrated in figure 6; once again, because two geodesics in D can intersect at most once, we see that certain cycles of symbols which arise at the vertices cannot occur. It turns out that, modulo a number of non-trivial mathematical difficulties which embody some very beautiful geometry (Morse 1938, and in the general case Series 1986), for a large class of fundamental regions, there is a finite list of excluded blocks, and that any symbol sequence containing none of the excluded blocks occurs as the symbol sequence of some trajectory. In the jargon, one says that the symbol sequences that occur are a 'subshift of finite type'.

The important, indeed crucial, point here is that sequences that occur as trajectories may be recognized as those satisfying certain simple rules, further, apart from a certain countable set, each trajectory on the surface corresponds to a *unique* symbol sequence. (Problems about trajectories that pass through vertices

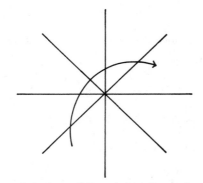

FIGURE 6. An impossible geodesic path round a vertex.

of R enter here.) Thus, to assert the existence of a geodesic with certain properties, it is enough to exhibit a symbolic trajectory with the analogous symbolic property. Geometry is recovered by the observation that symbol sequences that coincide over a block of length n correspond to trajectories that approach (in position and direction) within e^{-n}. For example, one easily obtains the following results, all characteristic of chaotic motion.

(i) Periodic sequences correspond to closed and periodic trajectories. The number of periodic trajectories increases exponentially with length. These trajectories can be systematically enumerated.

(ii) Periodic trajectories approach any possible trajectory arbitrarily closely.

More generally, the highly developed ergodic theory of subshifts of finite type applies to these systems and one obtains a wealth of results about their dynamical and metrical behaviour.

QUANTIZATION

There has been much interest recently in the problem of quantizing classically chaotic systems, c.f. e.g. Berry (1983). The problem is twofold: to find the correct substitute for the known procedures of quantizing classically chaotic systems, and to recognize those features of spectral behaviour that reflect chaotic motion.

It has been suggested by several authors, notably Gutzwiller (1983) and Balazs & Voros (1986), that one should return to the example of motion on hyperbolic surfaces to seek possible manifestations of quantum chaos. The motivation here is not the physical significance of the models; rather, these are examples in which the classical motion is completely chaotic and yet in which the mathematical machinery is sufficiently well developed that one can hope to obtain detailed results on the associated quantized systems. In what follows, I make no attempt to discuss physical interpretations, but confine myself to a few mathematical observations.

One seeks to solve the time-independent Schrödinger equation $\Delta f = -\lambda f$ in the Poincaré disc $\mathbb{D}$. In appropriate units, $\lambda = \hbar^{-2}$, where Δ is the non-euclidean laplacian operator given in geodesic polar coordinates t, θ, $z = re^{i\theta}$, $t = \mathrm{d}(0, z) = 2\,\mathrm{artanh}\,r$, by

$$\Delta = 1/(\sinh t)\,\partial/\partial t\,(\sinh t\,\partial/\partial t) + 1/(\sinh^2 t)\,\partial^2/\partial\theta^2.$$

Separating variables one obtains

$$f(z) = \sum_{m=-\infty}^{\infty} a_m\,\Phi_{\mu,\,m}(t)\,e^{im\theta}, \tag{1}$$

where $\Phi_{\mu,\,m}(t)$ is the associated Legendre function

$$\Phi_{\mu,\,m}(t) = 1/2\pi \int_0^{2\pi} (\cosh t - \sinh t \cos\theta)^{\mu}\,e^{im\theta}\,\mathrm{d}\theta \tag{2}$$

which satisfies $\Delta\Phi_{\mu,\,m} = \mu(\mu-1)\,\Phi_{\mu,\,m}$.

An argument that uses Green's theorem shows that

$$(\Delta f, f) \leqslant -\tfrac{1}{4}(f,f),$$

where $(,)$ is the non-euclidean inner product

$$(f,g) = 2/\pi \int_0^{2\pi} \int_0^1 f(r\,e^{i\theta})\,g(r\,e^{i\theta})\,r\,\mathrm{d}r\,\mathrm{d}\theta/(1-r^2)^2.$$

Thus $\mu(\mu-1) \leqslant \tfrac{1}{4}$, which implies that $\mu = \tfrac{1}{2} + i\rho$, $\rho \in \mathbb{R}$.

We now impose the further condition that f be invariant on the tesselation defining the surface in question; i.e. f is invariant under the group Γ generated by the side-pairing isometries Γ_R, $f(gz) = f(z)$, $g \in \Gamma$. This requirement of course imposes further conditions on μ and the coefficients a_m. One can regard these as boundary conditions: we seek f defined on the fundamental region R so that the values of f at boundary points identified by Γ coincide.

From general principles one knows that for a compact surface, μ can take on a discrete set of values, whereas for a non-compact surface one obtains both discrete and continuous spectrum. One may seek for manifestations of chaotic behaviour either in the spectral levels, or in the behaviour of individual eigenfunctions. Surprisingly little on either of these topics appears in the mathematical literature, and there seems to be much scope for further research. Here we simply collect a few known facts.

DISCRETE SPECTRUM

There are no known examples of explicit formulae for discrete eigenfunctions, and one is reduced to numerical studies. These are not easy. They have been carried out by number theorists, notably (Hejhal 1983, appendix C), for the modular group $\Gamma = SL(2, \mathbb{Z})$ (here the model for hyperbolic geometry is the upper halfplane $\mathbb{H}$ obtained from the unit disc $\mathbb{D}$ by the Cayley transform $z \to i(1+z)(1-z)^{-1}$); and by Schmit (1986) for the 'octagon group' associated to the fundamental region shown in figure 3a. In both cases one finds rather more clustering than would be predicted by random matrix theory; see (Balazs et al. 1987).

CONTINUOUS SPECTRUM

For surfaces with spikes (cusps) one has explicit formulae for the generalized eigenfunctions corresponding to the continuous spectrum. These are called Eisenstein series. If the group Γ contains only one cusp then the continuous spectrum is $\{-(\frac{1}{4}+\rho^2),\ \rho>0\}$ and the corresponding Eisenstein series is

$$E_\mu(z) = \Sigma_{\Gamma/\Gamma_\infty} \operatorname{Im}(gz)^\mu,\ \mu=\tfrac{1}{2}+\mathrm{i}\rho.$$

Here Γ is normalized to act on $\mathbb{H}$ with the cusp at ∞. Thus Γ contains a translation $T: z \to z+t$, $t \in \mathbb{R}$, and Γ_∞ is the subgroup $\{T^n: n \in \mathbb{Z}\}$ fixing ∞. The sum is taken over cosets Γ/Γ_∞ to avoid infinite repetition of identical terms. In the special case in which $\Gamma = SL(2,\mathbb{Z})$, this sum becomes the classical Eisenstein series $\Sigma_{(c,\,d)\,=\,1} \times 1/(cz+d)^\mu$. Notice that $f_\mu(z)=(\operatorname{Im}z)^\mu$ is an eigenfunction of $\varDelta=y^2(\partial^2/\partial x^2+\partial^2/\partial y^2)$, the non-euclidean laplacian in cartesian coordinates on $\mathbb{H}$. One can think of f_μ as a plane wave emanating from ∞, so that $E_\mu(z)$ is a superposition of plane waves emanating from all points equivalent to ∞ under Γ.

The Eisenstein series $E_\mu(z)$, being periodic under Γ, may be expanded as a Fourier series. For $\Gamma = SL(2,\mathbb{Z})$, generated by $T: z \to z+1$ and $J: z \to -1/z$, one obtains $E_\mu(z)=y^\mu+\varLambda(1-\mu)/(\varLambda(\mu))\,y^{1-\mu}+O(\mathrm{e}^{-2\pi y})$, where $y=\operatorname{Im}z$ and $\varLambda(\mu)=\pi^{-\mu}\varGamma(\mu)\,\zeta(2\mu)$. Here $\varGamma(\mu)$ is the Euler $\varGamma$-function and $\zeta(x)$ is the notorious Riemann zeta function defined as $\Sigma_{n\,=\,1}^\infty n^{-s}$ for $\operatorname{Re}s>1$ and analytically continued to the whole complex plane. Notice that we are interested in values of $\zeta(s)$ on $\operatorname{Re}s=1$; because of divergence of $\Sigma\,n^{-s}$ on this line, $\zeta(1+2\mathrm{i}\rho)$ cannot be evaluated directly from the sum.

This example has been related by Gutzwiller (1983) to the 'leaky torus' illustrated in figure 4b. This surface corresponds to a subgroup of index 6 in $SL(2,\mathbb{Z})$ and has a four sided fundamental region with all vertices on $\mathbb{R}\cup\{\infty\}$, figure 3c. Because $\varLambda$ is real on $\mathbb{R}$ one gets $\varLambda(1-\mu)\,\varLambda(\mu)^{-1}=\mathrm{e}^{\mathrm{i}\beta}$ for some $\beta\in\mathbb{R}$. This β can be interpreted as the phase shift of the reflected wave that results from an incoming wave y^μ. The interesting part of the behaviour of $\beta=\beta(\frac{1}{2}+\mathrm{i}\rho)$ is governed by that of $\zeta(s)$ on the line $\operatorname{Re}s=1$. This behaviour appears to be very chaotic, in that ζ approximates arbitrarily closely any non-vanishing analytic function in $\frac{1}{2}<\operatorname{Im}s<1$. More precisely (Voronin 1975; Reich 1980) it can be described by the following.

Let D be any disc in the region $\frac{1}{2}<\operatorname{Re}z<1$, $\operatorname{Im}z>0$. Let f be any function analytic and non-vanishing in D, and let $a>0$. Then the function $\zeta(z+\mathrm{i}na)$ approximates $f(z)$ arbitrarily closely on D (in the sup norm) for infinitely many $n\in\mathbb{N}$ (in fact, for a set of positive density in $\mathbb{N}$).

Further examples are obtained by replacing $SL(2,\mathbb{Z})$ by the Hecke groups G_n generated by $T_n: z \to z+2\cos\pi/n$ and $J: z \to -1/z$. For n odd, G_n has a subgroup of index $2n$ which has a $2n-2$ sided fundamental region all of whose vertices lie on $\mathbb{R}\cup\{\infty\}$, and which corresponds to a surface with n holes and one cusp. Here one obtains

$$E_\mu(z) = y^\mu + \psi_n(\mu)\,y^{1-\mu} + O(\mathrm{e}^{-2\pi y}).$$

For large n, a result of Selberg (Hejhal 1983, p. 579) shows that both zeros and poles of ψ_n occur arbitrarily close to every point on the critical line $\operatorname{Re}z=\frac{1}{2}$. Once again, one sees evidence of chaotic or at least very complicated behaviour.

The Helgason representation

Finally we turn to a representation of eigenfunctions f of Δ due to Helgason (1984) which we believe is useful in the study of the discrete spectrum. Helgason's method in this context has been discussed in some detail in Pignataro (1984).

Recall the classical Poisson representation of harmonic functions in $\mathbb{D}$: if $\Delta^{E} f = 0$ then

$$f(r\,e^{i\psi}) = \tfrac{1}{2}\pi \int_{0}^{2\pi} P(r, \theta - \psi) f(\theta)\,d\theta,$$

where Δ^{E} is the euclidean laplacian and $P(r, \theta) = (1 - r^2)(1 - 2r\cos\theta + r^2)^{-1}$ is the Poisson kernel.

The non-euclidean version of this, due to Helgason, is that if $\Delta^{H} f = -(\tfrac{1}{4} + \rho^2)f$, then

$$f(r\,e^{i\psi}) = \tfrac{1}{2}\pi \int_{0}^{2\pi} P(r, \theta - \psi)^{\frac{1}{2} + i\rho} f(\theta)\,d\theta,$$

where Δ^{H} is the non-euclidean laplacian as above. In fact $f(r\,e^{i\psi}) = P(r, \theta - \psi)^{\mu}$ is essentially the same basic wavefunction $(\operatorname{Im} z)^{\mu}$ as above, the difference being that we are now working in the disc $\mathbb{D}$ with the point ∞ replaced by $e^{i\theta}$. Thus the Helgason representation is simply a superposition of plane waves emanating from boundary points $e^{i\theta}$ with density $f(\theta)$.

Unfortunately, this representation is not, as it stands, mathematically rigorous, for we have implicitly assumed that the boundary values of f are given by an absolutely continuous density $f(\theta)\,d\theta$. We shall show that, for functions f invariant under a group Γ corresponding to a surface of finite area, such a representation is impossible.

The invariance condition $f(gz) = f(z)$, $z \in \Gamma$ implies that

$$f(g\theta) = \lg'(\theta)|^{\frac{1}{2} + i\rho} f(\theta). \tag{3}$$

Now this condition is too chaotic to be satisfied. In fact it is known that (Sullivan 1981):

given A, B $\subset \partial\mathbb{D}$, $t > 0$, $\epsilon > 0$; there exist A$' \subset$ A, B$' \subset$ B, $g \in \Gamma$ such that $g(A') = B'$, and

$$|g'(\theta)| - t| < \epsilon \quad \text{for} \quad e^{i\theta} \in A'.$$

(In the jargon, this is the statement that Γ acting on $\partial\mathbb{D}$ is type III, a fact 'well known' to the experts.) Brief inspection of (3) in the light of this remark should convince the reader that such a representation is impossible for any remotely reasonable function $f(\theta)$.

The correct statement is that $f(\theta)\,d\theta$ must be interpreted as a distribution T_f. Helgason has shown that such a representation is always possible provided that f satisfies a certain growth condition which is forced by our requirement $f(gz) = f(z)$. Those ρ for which condition (3) can be satisfied are precisely those ρ for which $-(\tfrac{1}{4} + \rho^2)$ is in the spectrum of Δ on $\mathbb{D}/\Gamma$.

It is not difficult to see that the Fourier coefficients $a_m = T_f(e^{-im\theta})$ are exactly the same as those occurring in the earlier expansion (1). In fact, one verifies easily that

$$\cosh t - \sinh t \cos \theta = P(r, \theta), \ r = \tanh \tfrac{1}{2}t,$$

so that (2) becomes

$$\Psi_{\mu, m}(r \, e^{i\psi}) = \Phi_{\mu, m}(t) \, e^{im\psi} = \tfrac{1}{2}\pi \int_0^{2\pi} P(r, \theta - \psi)^\mu \, e^{im\theta} \, d\theta, \qquad (4)$$

which incidentally exhibits the basic eigenfunctions $\Phi_{\mu, m}$ as superpositions of plane waves. From (4) we obtain immediately the Fourier expansion

$$P(r, \theta - \psi)^\mu = \sum_{m \in \mathbb{Z}} \Psi_{\mu, m}(r \, e^{i\theta}) \, e^{-im\theta}.$$

Applying T_f gives

$$f(r \, e^{i\psi}) = \sum a_m \, \Psi_{\mu, m}(r \, e^{i\psi}), \qquad (5)$$

and comparison with (1) gives

$$a_m = \Phi_{\mu, m}(t)^{-1} \tfrac{1}{2}\pi \int_0^{2\pi} f(r \, e^{i\theta}) \, e^{-im\theta} \, d\theta.$$

Voros has suggested the possibility that the behaviour of the a_n may be 'chaotic'. There is some asymptotic evidence in this direction, but much work remains to be done.

In conclusion, let me note that in the same circle of ideas, a recent remark of Pollicott (1986) indicates that the boundary distributions T_f may relate directly to symbolic dynamics on the surfaces via the theory of Gibbs distributions and resonances described by Ruelle (1986).

I am indebted to M. Gutzwiller, N. L. Balazs and A. Voros for many comments on this paper. I thank R. Fenn for drawing figure 3a, and S. Katok for drawing figures 3b, c and 5.

References

Artin, E. 1924 Ein mechanisches System mit quasiergodischen Bahnen. *Abh. Math. Sem., Hamburg*, vol. 3. (Collected papers, pp. 499–501. Reading Massachusetts: Addison Wesley (1965).)

Balazs, N. L. & Voros, A. 1986 Chaos on the pseudosphere. *Physics Rep.* **143**, 109.

Balazs, N. L., Schmit, C. & Voros, A. 1987 Spectral fluctuations and zeta functions. *J. statist Phys.* (In the press.)

Berry, M. V., 1983 In *Chaotic behaviour of deterministic systems, Proceedings 36, Les Houches 1981* (ed. G. Iooss et al.). Amsterdam: North Holland.

Bowen, R. 1973 Symbolic dynamics for hyperbolic flows. *Am. J. Math.* **95**, 429–460.

Gutzwiller, M. 1983 Stochastic behaviour in quantum scattering. *Physica* D. **7**, 341–355.

Hadamard, J. 1898 Les surfaces a coubures opposées et leurs lignes géodesiques. *J. de Math. pures appl.* **4**, 27–74.

Hedlund, G. 1935 A metrically transitive group defined by the modular group. *Am. J. Math.* **57**, 668–678.

Hedlund, G. 1934 On the metrical transitivity of geodesics on closed surfaces of constant negative curvature. *Ann. Math.* **35**, 787–808.

Hejhal, D. 1983 The Selberg trace formula for PSL(2, ℝ). *Springer Lecture notes in Mathematics 1001*, vol. 2.

Helgason, S. 1984 Groups and geometric analysis. New York: Academic Press.

Koebe, P. 1929 Riemannische Mannifaltigkeiten und nichteuklidische Raumformen. IV. *Sitzungberichte der Preussichen Akad. der Wissenschaften*, pp. 414–457.

Koebe, P. 1917 Urmanuskript der Preisschrift, deposited in Mittag-Leffler institute. (See *Acta Math.* **50**, 157 (1927).)

Morse, M. 1921 A one-to-one representation of geodesics on a surface of negative curvature. *Am. J. Math.* **43**, 33–51.

Morse, M. 1966 Symbolic dynamics. *Institute for Advanced Study Notes.* Princeton. (Unpublished notes written in 1938.)

Pignataro, T. 1984 Hausdorff dimension, spectral theory and applications to the quantization of geodesic flows on surfaces of constant negative curvature. Ph.D. thesis, University of Princeton.

Pollicott, M. 1986 Distributions at infinity for Riemann surfaces. *Proceedings of Special Semester on Dynamical Chaos, Stefan Banach Centre, Warsaw.*

Ratner, M. 1973 Markov partitions for Anosov flows on n-dimensional manifolds. *Israel J. Math.* **15**, 92–114.

Reich, A. 1980 Werteverteilung von Zetafunktionen. *Arch. Math.* **34**, 440–451.

Ruelle, D. 1986 Locating resonances for axiom A dynamical systems. Preprint.

Schmit, C. 1987 (In preparation.)

Series, C. 1986 Geometrical Markov coding of geodesics on surfaces of constant negative curvature. *Ergodic Theory Dynam. Sys.* **6**, 601–625.

Sinai, Ya. 1968 Markov partitions and C-diffeomorphisms. *Funct. Analysis* Appl. **2**, 64–89.

Sullivan, D. 1981 On the ergodic theory at infinity of an arbitrary discrete group of hyperbolic motions. *Ann. Math. Stud.* **97**, 465–496.

Voronin, S. M. 1975 A theorem on the 'universality' of the Riemann zeta function. *Izv. Akad. Nauk. SSSR Seria Mat.* **39**, 474–486. (In Russian.)

THE BAKERIAN LECTURE, 1987

Quantum chaology

By M. V. Berry, F.R.S.

H. H. Wills Physics Laboratory, Tyndall Avenue, Bristol BS8 1TL, U.K.

(*Lecture delivered* 5 *February* 1987 – *Typescript received* 2 *March* 1987)

Bounded or driven classical systems often exhibit chaos (exponential instability that persists), but their quantum counterparts do not. Nevertheless, there are new régimes of quantum behaviour that emerge in the semiclassical limit and depend on whether the classical orbits are regular or chaotic, and this motivates the following definition.

Definition. Quantum chaology is the study of semiclassical, but non-classical, behaviour characteristic of systems whose classical motion exhibits chaos.

This is illustrated by the statistics of energy levels. On scales comparable with the mean level spacing (of order h^N for N freedoms), these fall into universality classes: for classically chaotic systems, the statistics are those of random matrices (real symmetric or complex hermitian, depending on the presence or absence of time-reversal symmetry); for classically regular ones, the statistics are Poisson. On larger scales (of order h, i.e. classically small but semiclassically large), universality breaks down. These phenomena are being explained by representing spectra in terms of classical closed orbits: universal spectral behaviour has its origin in very long orbits; non-universal behaviour depends only on short ones.

In Henry Baker's day, 'chaology' meant 'The history or description of *the* chaos' (O.E.D. 1893). *The* chaos was the state of the world before creation ('without form, and void') so that chaology was a theological term. That area of theology has not been very active for the past two centuries (unless we extend its scope to include some recent speculations in cosmology) and so we are justified in reviving the term chaology, which will now refer to the study of unpredictable motion in systems with causal dynamics, as exemplified by the contributions at the meeting on 'dynamical chaos' of which this lecture is a part.

But what is 'quantum chaology'? One obstacle to a definition is the growing understanding that quantum systems are not chaotic in the way that classical systems are. (I am speaking of unpredictability in the evolution of the expectation values of observable quantities, and not of the quite different randomness unavoidably encoded in the wavefunction.)

As an example, consider ionizing a hydrogen atom by shining microwaves on it. This is well modelled by the quantum mechanics of an electron in two electric fields: Coulomb, from the nucleus, and oscillatory, from the radiation. If the atom is highly excited to begin with, we might be justified, on the basis of the correspondence principle, in thinking of the electron as moving classically. If in addition the illuminating microwaves are intense, the classical progress towards ionization is not a smooth outward spiralling but an erratic diffusion: the fields make the electron orbits chaotic (Leopold & Percival 1979; Jensen 1985). Exactly this behaviour (or rather the ionization probabilities that follow from it) has been observed in experiments (Bayfield et al. 1977). (We are here very far from the perturbation régime of one-photon ionization, the photoelectric effect, that was so important at the birth of quantum mechanics.) Surely these experiments illustrate 'quantum chaos'? They do not, because chaos is unpredictability that *persists* (strictly for infinite times) and in these experiments the atoms traverse only a short stretch of microwave field and so diffuse for only a short time.

The surprise comes in quantum calculations for longer times. These show that although initially the highly excited quantum electron absorbs energy in the classical way (that is, diffusively), after a long time there is a transition to a new régime in which the quantum electron absorbs energy more slowly. The first calculations (Casati et al. 1979) were for a model system, in which a particle on a ring (a rotator) is kicked periodically with an impulse that depends on where it is. For strong kicks the classical rotator momentum diffuses (energy grows linearly). But the quantum energy almost always eventually stops growing (usually it oscillates quasiperiodically). The analogous régime for the ionization problem (Casati et al. 1984; Casati et al. 1986; Blümel & Smilansky 1987) has not yet been probed experimentally, although I understand that it soon will be.

These calculations are important because they illustrate a general phenomenon: the quantum suppression of classical chaos (Chirikov et al. 1981; Fishman et al. 1982; Grempel et al. 1984). To see easily that this suppression must occur, observe first that classical chaos can be regarded as the emergence of complexity on infinitely fine scales in classical phase space: smooth curves representing families of orbits develop elaborate convolutions, like cream spreading on coffee. But quantum mechanics involves Planck's constant h, which is an area in phase space (momentum times distance) below which structure is smoothed away (for an illustration see Korsch & Berry 1981).

Although we do not have chaotic quantum evolution, we do have here a *new quantum phenomenon* that emerges in the semiclassical limit in systems that classically *are* chaotic, and this motivates the following definition.

Definition. Quantum chaology is the study of semiclassical, but non-classical, behaviour characteristic of systems whose classical motion exhibits chaos.

'Semiclassical' here means 'as $h \to 0$'. (Of course Planck's constant is not dimensionless and so can take any value, depending on the choice of units; what is meant is that the ratio of h to some classical quantity with the same dimensions – action – tends to zero.)

Here I will concentrate not on time evolution but on the quantum chaology of *spectra*, that is eigenvalues of the energy operator for systems whose classical counterparts are chaotic. This is important because these eigenvalues are the energies of stationary states, which are the quantum mechanical way of describing *things*, that is persisting objects like atoms and molecules, whose properties do not depend on when we measure them. We will be concerned not with the ground state but with the description of many highly excited states; this is the semiclassical limit.

My main aim is to bring to your attention a remarkable quantum chaotic property of spectra, and describe the first step towards explaining it. Before doing so, I must point out that these semiclassical quantum problems are but one example of the asymptotics of eigenvalues. Essentially the same mathematics describes the modes of vibration of elastic membranes, or sound waves (in a lecture hall my voice excites modes near the 20000th, which is surely 'asymptotic'), and much else. The 'classical limits' of these non-quantum problems involve the 'rays' of elasticity or sound; geometrically the rays are geodesics: straight line trajectories reflected specularly, like billiard balls, at the boundaries of the domain. The two-dimensional billiard domain of figure 1a has chaotic geodesics: it is the stadium of Bunimovich. The domain of figure 1b, the circle, does not. In mechanical terminology, the stadium orbits are *ergodic* (they possess no constants of motion other than the energy) while the circle orbits are *integrable* (because of symmetry, their angular momentum is conserved as well). For a quantum particle of mass m in a billiard domain D, eigenvalues E are determined by

$$\left. \begin{array}{l} \nabla^2 \psi + (2mE/\hbar^2)\psi = 0 \text{ in D,} \\ \psi = 0 \text{ on the boundary of D.} \end{array} \right\} \tag{1}$$

The remarkable quantum chaotic property is that the distribution of the eigenvalues displays *universality*. This is the slightly pretentious way in which physicists denote identical behaviour in different systems. The most familiar example is thermodynamics near critical points (of, say, fluids and magnets).

To see the universality we need to magnify the spectrum so that the mean spacing of the levels is unity. The required magnification is the mean level density $\langle d \rangle$. What is $\langle d \rangle$? The answer comes from the roughest eigenvalue asymptotics, initiated by Pockels in 1891, developed by Rayleigh and Jeans who needed to count cavity modes for the theory of black-body radiation, and given a firm mathematical foundation by Herman Weyl in 1913 (for a review see Baltes & Hilf 1976). Their result was that if the classical system has N freedoms (e.g. $N = 2$ for billiards) then

$$\langle d(E) \rangle \to \frac{d\Omega(E)/dE}{h^N} \quad \text{as} \quad h \to 0, \tag{2}$$

where $\Omega(E)$ is the volume of that part of classical phase space whose points have energies less than E. (These ideas have been refined and extended in several directions: see for example Kac 1966; Simon 1983a,b; Berry 1987.)

The level spacing is thus of order h^N, so that we need a microscope with power h^{-N}. What do we see with it? Of course we see the individual scaled levels, call them x_j, instead of the original levels E_j. Ideally we would have an asymptotic theory to predict these levels with an error that gets semiclassically small in comparison with the mean spacing h^N. For integrable (non-chaotic) classical motion we do have such a theory, in the form of the W.K.B. method and its

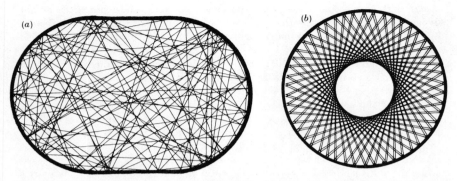

FIGURE 1. Classical orbits (bouncing geodesics) in billiards: (a) stadium of Bunimovich (chaotic), (b) circle (regular). For more details see, for example, Berry 1981 a.

refinements and descendants (see, for example, Berry & Mount 1972; Percival 1977; Berry 1983). And these methods can be extended far into the chaotic régime if there is some residual order in phase space ('vague tori') and under not too semiclassical conditions (Reinhardt & Dana, this symposium). But for fully chaotic systems no fully asymptotic eigenvalue theory exists: we must make do with *statistics* of levels, and it is these that exhibit universality.

One such statistic, a short-range one, is the *level spacings probability distribution* $P(S)$, that is, the distribution of $S_j = x_{j+1} - x_j$. Figure 2 shows $P(S)$ computed from several hundred levels of the stadium, superimposed on $P(S)$ for another

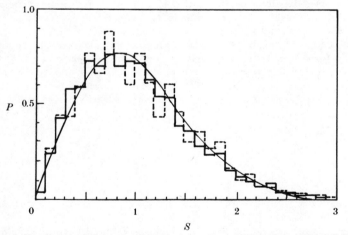

FIGURE 2. Level spacing histograms $P(S)$ for eigenvalues of the stadium billiard (full lines, after Bohigas 1984a) and the Sinai billiard (dashed lines, after Bohigas *et al.* 1984b), and the level spacings distribution for random real symmetric matrices (smooth curve), closely approximated by $P(S) = (\frac{1}{2}\pi \exp\{-\frac{1}{4}\pi S^2\}$.

classically chaotic billiard system: the billiard of Sinai, which is a square with a circular obstacle at its centre. These are two different systems, but the distributions are evidently the same; this is universality.

Another statistic – a long-range one – is the *spectral rigidity* $\Delta(L)$. This measures the fluctuations of the spectral staircase $\mathcal{N}(x)$, whose treads are at the eigenvalues x_j and whose risers have unit height ($\mathcal{N}(x)$ counts the number of levels below x). The rigidity (Dyson & Mehta 1963) is the mean-square deviation of the staircase from the straight line that best fits it over a range L, that is

$$\Delta(L) = \left\langle \min_{A,B} \frac{1}{L} \int_{-\frac{1}{2}L}^{\frac{1}{2}L} L \, dx \, [\mathcal{N}(x) - Ax - B]^2 \right\rangle. \tag{3}$$

Figure 3 shows the rigidities for the same two chaotic billiards; again they are the same, illustrating universality.

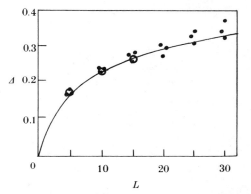

FIGURE 3. Spectral rigidity $\Delta(L)$ for eigenvalues of the stadium billiard (filled circles, after Bohigas *et al.* 1984*a*) and the Sinai billiard (open circles, after Bohigas *et al.* 1984*b*), and the rigidity for random real symmetric matrices (smooth curve), whose asymptote is $\Delta(L) \to (1/\pi^2) \ln L + \text{const.}$ as $L \to \infty$.

Now, it is clear from figures 2 and 3 that these data are accurately fitted by smooth curves representing the eigenvalue statistics of infinite real symmetric *matrices whose elements are random numbers*. Random-matrix theory (Porter 1965) was developed in the 1960s to model the complicated many-body energy operators for atomic nuclei (whose observed spectra they describe very well (Haq *et al.* 1982)). Ten years ago we (Berry & Tabor 1977) began to suspect it might also describe systems which although simple (like billiards) have chaotic classical orbits, and this has turned out to be so (Bohigas & Giannoni 1984).

Contrast this universality class with the spectral statistics of systems whose classical motion is not chaotic. Figure 4*a* shows the spacings distribution, and figure 4*b* the rigidity, for that most humble of regular systems, the particle in a two-dimensional rectangular box. It was surprising (ten years ago) to predict (Berry & Tabor 1977) and then find the statistics to be those of a set of *random*

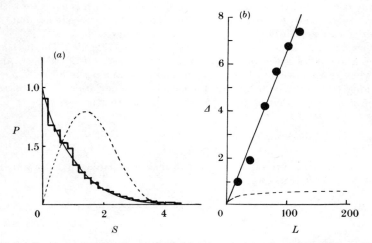

FIGURE 4. Level spacings distribution $P(S)$ (histogram in (a) and spectral rigidity $\Delta(L)$ (circles in (b)) for eigenvalues of the rectangular billiard. The full curves are the statistics for Poisson-distributed eigenvalues ($P(S) = \exp(-S)$ and $\Delta(L) = \frac{1}{15}L$) and the dashed curves are the statistics for random real symmetric matrices.

numbers (that is, poissonian). These behave very differently from the eigenvalues of random matrices, which are more well ordered in that they repel each other: for example, $P(S)$ vanishes linearly as $S \to 0$ instead of tending to a constant, and the asymptote of $\Delta(L)$ rise only logarithmically rather than linearly.

So far we have two universality classes, one for classically chaotic systems and one for classically regular systems, with spectra generated by random real symmetric matrices and Poisson processes respectively. Now, the matrices of quantum mechanics need not be real symmetric. The most general case is achieved for systems which, unlike billiards (or, more generally, particles in scalar potentials), *do not possess time-reversal symmetry* (T). For these, the energy operators are represented by complex hermitian, rather than real symmetric, matrices. The spectra of such random matrices, and also of the corresponding quantized chaotic systems, fall into a *third universality class*.

To illustrate it we break T by applying an external magnetic field to a charged particle moving chaotically. It is very instructive to concentrate the field into a single line of magnetic flux Φ. This is the chaotic equivalent of the effect discovered nearly thirty years ago in Bristol by Aharonov & Bohm (1959): the flux line does not alter the classical trajectories but does affect the quantum mechanics, in this case by changing the eigenvalues (Berry & Robnik 1986a). These are determined not by (1) but by

$$\left.\begin{array}{c} (\nabla - \mathrm{i}q\boldsymbol{A}(\boldsymbol{r})/\hbar)^2\psi + (2mE/\hbar^2)\psi = 0 \text{ in D}, \\ \psi = 0 \quad \text{on the boundary of D}, \end{array}\right\} \tag{4}$$

where $\boldsymbol{A}(\boldsymbol{r})$ is any vector potential satisfying $\nabla \times \boldsymbol{A} = \Phi\delta(\boldsymbol{r})$.

Figure 5 shows the spectral statistics of an Aharonov–Bohm billiard ('Africa') with chaotic trajectories (Africa is a cubic conformal image of the unit disc, illustrated in figure 6). Evidently $P(S)$ now vanishes quadratically as $S \to 0$, rather than linearly. The rigidity is different too: its logarithmic asymptote is only half that for chaotic systems with T. Thus T-breaking induces a *spectral phase transition*, to the third universality class. (Additional symmetries can mimic the effect of T, as explained by Robnik & Berry 1986). The Aharonov–Bohm chaotic billiard might appear contrived, but might be capable of realization with a tiny solenoid and the essentially two-dimensional electrons in certain semiconductor interfaces (M. Pepper, personal communication). Exact sum rules for Aharonov–Bohm eigenvalues are given by Berry (1986 a).

It is instructive to digress and look at the *wavefunctions* of these systems without T, and particularly at their zeros (Berry & Robnik 1986 b). *With T*, wavefunctions

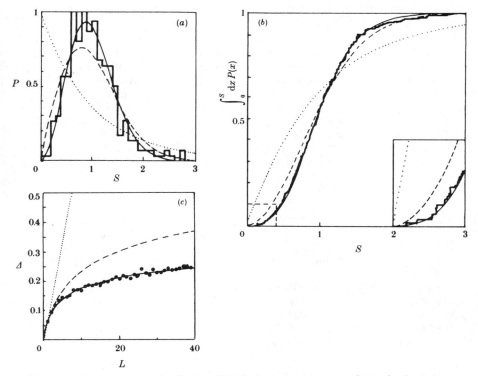

FIGURE 5. Level spacings distribution $P(S)$ (histogram in (a)), cumulative level spacings distribution $\int_0^S \mathrm{d}x P(x)$ (histogram in (b)), and spectral rigidity $\varDelta(L)$ (circles in (c)) for eigenvalues of the Aharonov–Bohm 'Africa' billiard with flux $q\varPhi/h = \frac{1}{2}(\sqrt{5}-1)$. The full curves are the statistics for random complex hermitian matrices, for which $P(S) \approx (32/\pi^2)\exp(-4S^2/\pi)$ and $\varDelta(L) \to (1/2\pi^2)\ln L + \mathrm{const.}$ as $L \to \infty$. The dashed curves are the statistics for random real symmetric matrices and the dotted curves are Poisson statistics.

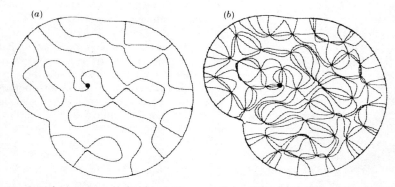

FIGURE 6. 50th eigenstate ψ of 'Africa' Aharonov–Bohm billiard with flux $q\Phi/h = \frac{1}{2}(\sqrt{5}-1)$. (a) Nodal lines of Re ψ, which are very similar to the nodal lines of ψ with zero flux; (b) wavefronts (contours of phase of ψ) at intervals of $\frac{1}{4}\pi$.

are real and so in two dimensions their zeros are the familiar *nodal lines* ($\psi(x,y) = 0$), which in quantum chaotic systems wander irregularly (figure 6a) with average spacing equal to the de Broglie wavelength $\lambda = \hbar/\sqrt{2mE}$ (McDonald & Kaufman 1979; Berry 1983; see also Heller 1984, 1986). Without T, wavefunctions are inescapably complex and so their zeros are points (Re $\psi(x,y) = 0$, Im $\psi(x,y) = 0$). Each of these points is a singularity of the wavefronts (contours of the phase of ψ) (figure 6b), which radiate from it like spokes from an axle (Nye & Berry 1974; Berry 1981b). Classical waves, like those on the surface of the sea, are of course real, but share the properties of 'inescapably complex' ones if their patterns are stationary but not standing, that is Re ψ, where

$$\psi(\boldsymbol{r}, t) = F(\boldsymbol{r}) \, \mathrm{e}^{-\mathrm{i}wt} \tag{5}$$

with $F(\boldsymbol{r})$ complex; thus the nodal lines of Re ψ move. The *tide waves* are like this, because of the symmetry-breaking caused by the Earth's rotation (relative to the Moon), and the phase singularities are the *amphidromic points* where the cotidal lines (wavefronts) meet (figure 7), as described by Whewell (1833, 1836) (see also Defant 1961).

Back now to eigenvalues. There is a set of numbers of great mathematical importance whose statistics precisely mimic the energy levels of a quantum chaotic system without T, namely the imaginary parts of the *zeros of Riemann's zeta function*. This function is defined (Edwards 1974) by analytically continuing to the whole complex z-plane Euler's product over primes p:

$$\zeta(z) = \prod_{p} \frac{1}{1-p^{-z}}. \tag{6}$$

Riemann showed that the zeros of $\zeta(z)$ determine the fluctuations in the density of primes (that is their importance) and conjectured that they all have real part $\frac{1}{2}$; thus

$$\zeta(\tfrac{1}{2}+\mathrm{i}E_j) = 0, \tag{7}$$

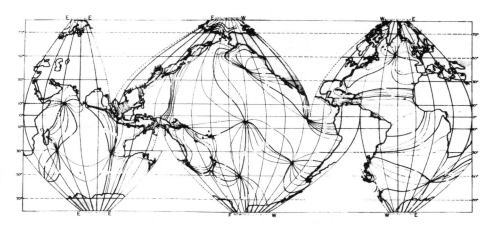

FIGURE 7. Cotidal lines in the oceans. These are wavefronts of the 12h tide wave, a forced
vibration of the water of the whole earth; each line connects points where the tide is high
at a given time. The singularities are amphidromic points, where there is no tide. (From
Defant 1961.)

where $\{E_j\}$ are real. This conjecture has been verified by computation for the first
1.5×10^9 zeros (Van de Lune *et al.* 1986). It is an old idea (going back at least to
Hilbert & Polya) that the Riemann conjecture would be confirmed if it could be
shown that $\{E_j\}$ are the eigenvalues of some hermitian operator, but this has not
been found.

Recently Odlyzko (1987) has computed some statistics for spectacularly high
E_j. Figure 8a shows the spacings distribution for 10^5 zeros near the 10^{12}th; agreeing
very closely with $P(S)$ for random complex hermitian matrices and so with that of
some unknown quantum system without T whose unknown classical limit is
chaotic. He also computed the number variance (figure 8b), a quantity closely
related to the rigidity, and discovered that the three- and four-zero correlations
(figure 8c, d) agrees perfectly with the corresponding complex random-matrix
statistics. Riemann's conjecture thus acquires, in addition to its number-theoretic
importance, a further significance (Berry 1986b): when (if) the operator with
eigenvalues E_j is found, it will surely be simple, and will provide a paradigm for
quantum chaology comparable with the harmonic oscillator for quantum non-
chaology.

Here is a way of breaking T without magnetic fields, in *relativistic quantum
chaology*. Take a massless particle ('neutrino') moving in the plane and described
by the equation of Dirac (who gave this lecture in the year of my birth), but with
a four-scalar potential $V(x,y)$ rather than the usual electric potential. For such a
particle, the wave is a two-component spinor satisfying (Berry & Mondragon
1987)

$$\begin{pmatrix} V(x,y) & -ihc(\partial_x - i\partial_y) \\ -ihc(\partial_x + i\partial_y) & -V(x,y) \end{pmatrix} \begin{pmatrix} \psi_1 \\ \psi_2 \end{pmatrix} = E \begin{pmatrix} \psi_1 \\ \psi_2 \end{pmatrix}. \tag{8}$$

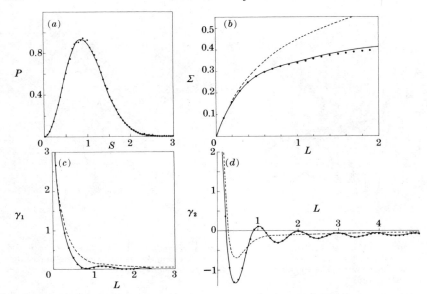

FIGURE 8. Statistics of imaginary parts of Riemann zeros, computed by A. M. Odlyzko. (a) $P(S)$
(from Odlyzko 1987); (b) number variance $\Sigma(L) \equiv \langle (n-L)^2 \rangle$, where n is the actual number
of zeros in an interval where the average number is L (the interval is $2\pi L / \ln(E/2\pi e)$) (from
data kindly supplied by A. M. Odlyzko); (c) Skewness $\gamma_1(L) \equiv \langle (n-L)^3 \rangle / \langle (n-L)^2 \rangle^{\frac{3}{2}}$
(kindly supplied by O. Bohigas); (d) excess $\gamma_2(L) \equiv \langle (n-L)^4 \rangle / \langle (n-L)^2 \rangle^2 - 3$ (kindly
supplied by O. Bohigas). Full curves, random complex hermitian matrices; dashed curves,
random real symmetric matrices.

This equation does not possess time-reversal invariance. Figure 9 shows the
spectral statistics when $V(x, y)$ represents a hard wall (neutrino billiards), showing
once again the statistics of complex hermitian random matrices if the billiard is
classically chaotic, and Poisson statistics if it is regular.

Originally I hoped, following a suggestion of Professor Atiyah, that this kind of
relativistic quantum chaology might help in the search for the elusive Riemann
operator, but this has not yet proved to be so. However, Volkov & Pankratov
(1985) and Pankratov *et al.* (1987) have recently discovered that an equation very
similar to (7) appears to describe peculiar electron states localized in the interface
between certain pairs of semiconductors (e.g. PbTe and SnTe, and HgTe and
CdTe).

There is a *fourth* universality class, associated with chaotic systems that have
time-reversal symmetry and also half-integer total spin (Porter 1965), but I will
not speak about it.

So far we have seen that on fine scales the statistics of spectra fall into universality classes that depend on whether the classical motion is regular or chaotic,
and on the symmetry of the energy operator. Now I have to explain how this
universality is compromised in two ways.

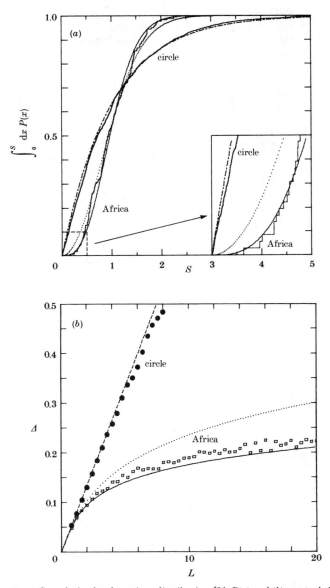

FIGURE 9. (a) Cumulative level spacings distribution $\int_0^S \mathrm{d}x P(x)$ and (b) spectral rigidity $\Delta(L)$, for neutrino 'Africa' and neutrino circle billiards. Full curves, random complex hermitian matrices; dotted curves, random real symmetric matrices; dashed and chain curves, Poisson statistics.

First, some very important systems are partly regular and partly chaotic in their classical motion; vibrating molecules, for example. Their spectral statistics can be understood as those of a *superposition* of spectra from different universality classes, each spectrum being associated with a different chaotic or regular region in classical phase space (Berry & Robnik 1984). Figure 10 shows some recent calculations by Wunner *et al.* (1986), of the spacings distribution of the zero-angular-momentum, even-parity electron levels of a hydrogen atom in a very strong magnetic field (6T), in three different energy ranges. The point is that the

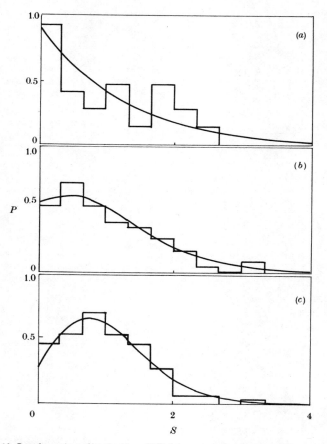

FIGURE 10. Level spacings distributions $P(S)$ for even-parity, zero-angular momentum energy levels of a hydrogen atom in a 6T uniform magnetic field, for three different energy ranges with different phase-space fractions q of regular orbits: (a) $-130 \text{ cm}^{-1} < E < -100 \text{ cm}^{-1}$ ($q = 0.71$, 47 levels); (b) $-100 \text{ cm}^{-1} < E < -70 \text{ cm}^{-1}$ ($q = 0.32$, 71 levels); (c) $-70 \text{ cm}^{-1} < E < -40 \text{ cm}^{-1}$ ($q = 0.16$, 116 levels); the smooth curves are $P(S)$ for superpositions of Poisson and random real symmetric matrix spectra. (From Wunner *et al.* 1986.)

corresponding classical motion gets more chaotic as the energy increases. These régimes are now within the reach of experiment (and are of course far removed from the familiar low-field 'perturbation' domain of the Zeeman effect).

The second compromise, of deep theoretical significance, is that universality is only local: for correlations involving very many levels, it breaks down. Recall the h^{-N} microscope that magnified the energies E_j to the numbers x_j with mean spacing unity, and note that N, the number of classical freedoms, is at least two for non-trivial cases. Now reduce the microscope's power to h^{-1}. (These gedankenmagnifications are strongly reminiscent of the 'non-standard analysis' used nowadays to describe infinitesimals (Harnik 1986).) We will see energy ranges that are still *classically* small (of order h) but *semiclassically large* in that they include many levels (a number of order $h^{-(N-1)}$). At these magnifications, energy-level statistics are not universal: they depend on classical details.

To illustrate the breakdown of universality at long range, figure 11*a* shows the rigidity for the (classically regular) particle in a rectangular box, computed by Casati *et al.* (1985). When L is not too large we see the straight line of the 'universal' Poisson statistics (this was figure 4*b*), but when L approaches the square root of the number of the highest level included in the calculation (which for this case corresponds to an energy range of order h), $\Delta(L)$ oscillates and then saturates at a value that depends on this number and also on the aspect ratio of the rectangle; that is, non-universally. Figure 11*b* shows the number variance for the Riemann zeros (underlying which there appears to be a chaotic classical system). When L is not too large we see the logarithmic curve of the 'universal' statistic for random complex matrices (this was figure 8*b*), but for larger L the variance oscillates about a value which depends on the number of zeros, that is, non-universally.

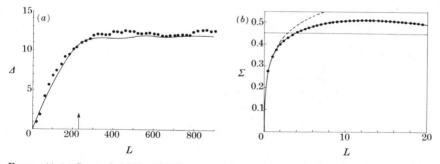

FIGURE 11. (a) Spectral rigidity $\Delta(L)$ for rectangular billiard, continuing figure 4*b* to larger L. The circles were computed from the eigenvalues near the 20000th Casati *et al.* 1985); the smooth curve was obtained from the sum over closed orbits by Berry (1985); the arrow is the L corresponding to an energy range h/T_{min}, where T_{min} is the period of the shortest closed orbit. (b) Number variance $\Sigma(L)$ for 10^5 Riemann zeros near the 10^{12}th, continuing figure 8*b* to larger L; the circles are plotted from data kindly supplied by A. M. Odlyzko; the smooth curve is the 'semiclassical' theory (adapted from Berry 1985), which predicts oscillation about the horizontal line $\Sigma(\infty) = 0.4518$ ($= [\ln \ln (E/2\pi) + 1.2615]/\pi^2$, where $(E/2\pi) \ln (E/2\pi e) = 10^{12}$); the dashed curve is for random complex matrices.

Until now I have spoken of spectral universality as an unexplained observation based on numerical experiments inspired by guesses. And so it was until recently, but now we have the beginnings of a theory (Berry 1985). Because of the h-magnifications involved, the theory has to be semiclassical: we must 'sew the quantum flesh on the classical bones'.

What are these bones? According to a beautiful picture developed by Gutzwiller (1971, 1978) and by Balian & Bloch (1972), they are the *classical closed orbits*, in terms of which an asymptotic formula can be given for the density of quantum eigenvalues (for a review, see Berry 1983; for the application to integrable systems, see Berry & Tabor 1976). These ideas can be traced back to de Broglie who in 1923 conceived of quantization as the constructive self-inteference of waves accompanying orbiting particles (think of Ouroboros, the mythical self-swallowing snake). For some mathematical systems (the Laplace–Beltrami operator on surfaces of constant negative curvature), the relation between spectra and closed geodesics is exact rather than asymptotic, and is called the Selberg trace formula (McKean 1972; Hejhal 1976; Balazs & Voros 1986; Series, this symposium.)

With the exception of some simple cases, the quantum levels are not in one-to-one correspondence with closed orbits (for an illustration, see Keating & Berry 1988; if they were, we would have a general formula for semiclassical quantization). Instead, each classical orbit describes an *oscillatory clustering* of the levels on a scale ΔE determined by its period T_n: this scale is just what would be expected from the uncertainty principle:

$$\Delta E = h/T_n. \tag{9}$$

Thus longer orbits give spectral information on finer scales, and it is this observation that gives the key to understanding the universality of the statistics (Berry 1985). With the h^{-N} microscope, we are concerned with the finest scales of spectral structure, of the order of the mean level spacing, so $\Delta E \sim h^N$. These scales depend on classical orbits with periods $T_n \sim h/\Delta E \sim 1/h^{(N-1)}$, that is, on *extremely long orbits*. Now, the distribution of these long orbits in phase space is very different for integrable and chaotic systems. For integrable systems, the orbits form continuous families whose number grows with period as T^N. For chaotic systems, the orbits are isolated and unstable and their number proliferates exponentially (as $\exp(HT)/HT$ where H is the Kolmogorov entropy – instability exponent of the orbit). In an important paper, Hannay & Ozorio de Almeida (1984) have shown that the way these long orbits contribute to one form of the asymptotic spectral formula is *universal*: it depends only on whether the orbits are chaotic or not, and on no other feature of the classical motion. It is this *classical* universality that begets the *quantum* universality, for it is possible to employ it as one ingredient in a derivation (Berry 1985) of the spectral rigidity $\Delta(L)$ (but not, so far, the spacings distribution), yielding precisely the Poisson and random-matrix formulae that so accurately fit the numerical computations.

The same arguments explain why universality breaks down at the larger energy scales $\Delta E \sim h$: the spectral fluctuations in this range are determined by orbits with period $T \sim h/\Delta E \sim h^0$, which are not long and so differ from system to system. The quantitative theory of this breakdown of universality (Berry 1985) works rather well, as figure 11 shows.

In summary, the vigorous development of quantum chaology during the last decade has been stimulated by the interplay of two factors: the realization that chaotic motion is ubiquitous in classical mechanics, and the discovery of associated new régimes of quantum behaviour. But the mathematical difficulties in understanding these régimes are severe, fundamentally because the semiclassical limit $h \to 0$ is highly singular. At the risk of sounding slightly paradoxical, I would say that we are discovering the connections between classical mechanics and quantum mechanics to be richer and more subtle than either mechanics is when considered on its own.

REFERENCES

Aharonov, Y. & Bohm, D. 1959 *Phys. Rev.* **115**, 485–491.
Balazs, N. L. & Voros, A. 1986 *Physics Rep.* **143**, 109–240.
Balian, R. & Bloch, C. 1972 *Ann. Phys.* **69**, 76–160.
Baltes, H. P. & Hilf, E. R. 1976 *Spectra of finite systems.* Mannheim: B. I. Wissenschaftsverlag.
Bayfield, J. E., Gardner, L. D. & Koch, P. M. 1977 *Phys. Rev. Lett.* **39**, 76–79.
Berry, M. V. 1981*a* *Eur. J. Phys.* **2**, 91–102.
Berry, M. V. 1981*b* Singularities in waves and rays. In *Physics of defects* (*Les Houches Lectures* 34) (ed. R. Balian, M. Kleman & J.-P. Poirier), pp. 453–543. Amsterdam: North Holland.
Berry, M. V. 1983 Semiclassical mechanics of regular and irregular motion. In *Chaotic behaviour of deterministic systems* (*Les Houches Lectures* 36) (ed. G. Iooss, R. H. G. Helleman & R. Stora), pp. 171–271. Amsterdam: North Holland.
Berry, M. V. 1985 *Proc. R. Soc. Lond.* A **400**, 229–251.
Berry, M. V. 1986*a* *J. Phys.* A **19**, 2281–2296.
Berry, M. V. 1986*b* In *Quantum chaos and statistical nuclear physics* (ed. T. H. Seligman & H. Nishioka) (Springer lecture notes in physics no. 263), pp. 1–17.
Berry, M. V. 1987 *J. Phys.* A **20**, 2389–2403.
Berry, M. V. & Mondragon, R. J. 1987 *Proc. R. Soc. Lond.* A **412**, 53–74.
Berry, M. V. & Mount, K. E. 1972 *Rep. Prog. Phys.* **35**, 315–397.
Berry, M. V. & Robnik, M. 1984 *J. Phys.* A **17**, 2413–2421.
Berry, M. V. & Robnik, M. 1986*a* *J. Phys.* A **19**, 649–668.
Berry, M. V. & Robnik, M. 1986*b* *J. Phys.* A **19**, 1365–1372.
Berry, M. V. & Tabor, M. 1976 *Proc. R. Soc. Lond.* A **349**, 101–123.
Berry, M. V. & Tabor, M. 1977 *Proc. R. Soc. Lond.* A **356**, 375–394.
Blümel, R. & Smilansky, U. 1987 *Z. Phys.* (In the press.)
Bohigas, O. & Giannoni, M. J. 1984 Chaotic motion and random-matrix theories. In *Mathematical and computational methods in nuclear physics* (ed. J. S. Dehesa, J. M. G. Gomez & A. Polls) (Lecture Notes in Physics 209), pp. 1–99. New York: Springer Verlag.
Bohigas, O., Giannoni, M. J. & Schmit, C. 1984*a* *J. Phys. Lett.* **45**, L1015–L1022.
Bohigas, O., Giannoni, M. J. & Schmit, C. 1984*b* *Phys. Rev. Lett* **52**, 1–4.
Casati, G., Chirikov, B. V., Ford, J. & Izraelev, F. M. 1979 In *Stochastic behaviour in classical and quantum hamiltonian systems* (ed. G. Casati & J. Ford) (Springer lecture notes in physics 93), pp. 334–352.
Casati, G., Chirikov, B. V. & Guarneri, I. 1985 *Phys. Rev. Lett.* **54**, 1350–1353.
Casati, G., Chirikov, B. V. & Shepelyansky, D. L. 1984 *Phys. Rev. Lett.* **53**, 2525–2528.
Casati, G., Chirikov, B. V., Shepelyansky, D. L. & Guarneri, I. 1986 *Phys. Rev. Lett.* **57**, 823–826.
Chirikov, B. V., Izraelev, F. M. & Shepelyansky, D. L. 1981 *Soviet Sci. Rev.* C **2**, 209–267.
Defant, A. 1961 *Physical oceanography*, vol. 2. London: Pergamon.
Dyson, F. J. & Mehta, M. L. 1963 *J. math. Phys.* **4**, 701–712.
Edwards, H. M. 1974 *Riemann's zeta function*, New York and London: Academic Press.
Fishman, Shmuel, Grempel, D. R. & Prange, R. E. 1982 *Phys. Rev. Lett.* **49**, 509–512.
Grempel, D. R., Fishman, Shmuel & Prange, R. E. 1984 *Phys. Rev.* A **29**, 1639–1647.
Gutzwiller, M. C. 1971 *J. math. Phys.* **12**, 343–358.

Gutzwiller, M. C. 1978 In *Path integrals and their applications in quantum statistical and solid-state physics* (ed. G. J. Papadopoulos & J. T. Devreese), pp. 163–200. New York: Plenum.

Hannay, J. H. & Ozorio de Almeida, A. M. 1984 *J. Phys.* A **17**, 3429–3440.

Haq, R. U., Pandey, A. & Bohigas, O. 1982 *Phys. Rev. Lett.* **48**, 1086–1089.

Harnik, V. 1986 *Math. Intell.* **8**, 41–47, 63.

Hejhal, D. A. 1976 *Duke math. J* **43**, 441–482.

Heller, E. J. 1984 *Phys. Rev. Lett.* **53**, 1515–1518.

Heller, E. J. 1986 In *Quantum chaos and statistical nuclear physics* (ed. T. H. Seligman & H. Nishioka) (Springer lecture notes in physics no. 263), pp. 162–181.

Jensen, R. V. 1985 In *Chaotic behaviour in quantum systems* (ed. G. Casati), pp. 171–186. New York: Plenum.

Kac, M. 1966 *Am. Math. Mon.* **73**, no. 4, part II, 1–23.

Keating, J. P. & Berry, M. V. 1988 (In preparation.)

Korsch, H. J. & Berry, M. V. 1981 *Physical* **3D**, 627–636.

Leopold, J. G. & Percival, I. C. 1979 *J. Phys.* B **12**, 709–721.

McDonald, S. W. & Kaufman, A. N. 1979 *Phys. Rev. Lett.* **42**, 1189–1191.

McKean, H. P. 1972 *Communs pure appl. Math.* **25**, 225–246.

Nye, J. F. & Berry, M. V. 1974 *Proc. R. Soc. Lond.* A **336**, 165–90.

Odlyzko, A. M. 1987 *Maths Comput.* **48**, no. 4, 273–308.

O.E.D. 1893 *A new English dictionary on historical principles*, vol. 2 (ed. J. A. H. Murray). Oxford: Clarendon Press.

Pankratov, O. A., Pakhomov, S. V. & Volkov, B. A. 1987 *Solid St. Commns* **61**, 93–96.

Percival, I. C. 1977 *Adv. chem. Phys.* **36**, 1–61.

Porter, C. E. 1965 *Statistical theories of spectra: fluctuations*, New York: Academic Press.

Robnik, M. & Berry, M. V. 1986 *J. Phys.* A **19**, 669–682.

Simon, B. 1983a *J. funct. Anal.* **53**, 84–98.

Simon, B. 1983b *Ann. Phys.* **146**, 209–220.

Van de Lune, J., te Riele, H. J. J. & Winter, D. T. 1986 *Maths Comput.* **46**, no. 74, 667–681.

Volkov, B. A. & Pankratov, O. A. 1985 *JETP Lett* **42**, 178–181.

Whewell, W. 1833 *Phil. Trans. R. Soc. Lond.* **123**, 147–236.

Whewell, W. 1836 *Phil. Trans. Roy. Soc. Lond.* **126**, 289–307.

Wunner, G., Woelk, U., Zech, I., Zeller, G., Ertl, T., Geyer, F., Schweizer, W. & Ruder, H. 1986 *Phys. Rev. Lett.* **57**, 3261–3264.

General discussion

[*Questioner not identified.*] Are there any applications of chaos to psychology?

E. C. ZEEMAN, It is advisable to be cautious about using chaos to model biological systems, particularly human behaviour, because there are usually so many influencing factors that it is difficult to isolate any part of the system for sufficiently long to allow a chaotic deterministic model to be tested. There is one part of the brain, however, where chaotic modelling may prove useful and that is the limbic brain.

Broadly speaking the human forebrain is divided into three layers, and, although there are strong pathways between the layers, nevertheless each layer tends to act as a separate unit anatomically, histologically, dynamically and functionally. Following MacLean, and again very broadly speaking, the top layer is the neocortex where language is stored and rational thinking occurs; the middle layer is the limbic brain where emotions and moods are generated; and the bottom layer is the R-complex, including the corpus striatus, where instincts are stored. We are aware of these three simultaneous activities in our minds most of the time.

Because the layers behave differently it is appropriate to use different branches of mathematics to model their activities. The top layer responds immediately to stimuli, and hence it is reasonable to use the minima of risk functions, the maxima of weighted sums, combinatorial and computer programmes, and indeed ordinary language. The middle layer exhibits striking hysteresis, because moods can often delay before responding to stimuli, and hence it is reasonable to use differential equations or high-dimensional dynamical systems to model the emotions. Each mood would be represented by its characteristic underlying oscillatory attractor, and a sudden switch of moods would be represented by the breakdown in stability of that attractor and a catastrophic switch to another.

There is one disorder, however, that might be modelled by a chaotic attractor, and that is schizophrenia. It is known that schizophrenia is accompanied by an increase in the number of dopamine receptors in the limbic brain, and sometimes displays EEC patterns resembling epilepsy, both of which features provide evidence for abnormal dynamics. Victims report 'whirling thoughts' and 'transient and inappropriate emotions'. Indeed the name schizophrenia originally referred to the apparent split between thoughts (neocortex) and emotions (limbic), which would certainly be observed if the limbic brain went into chaotic oscillations. Such a chaotic oscillator could well orbit through large regions of the mood-generating space, causing the victim to cry, laugh, rage and fear in quick succession, and as a result feel thoroughly bewildered and anxious.

A genetic tendency to increase the histological connectivity of this layer would provide a simple explanation of why the disorder occurs with steady probability, independent of environment and culture; and the resulting chaotic dynamics would generate the complicated and diffuse symptoms that are observed, and which may be difficult to explain without a thorough understanding of the theory of chaos.

[199]

Library of Congress Cataloging-in-Publication Data

Royal Society (Great Britain). Discussion Meeting (1987 : London, England)
Dynamical chaos : proceedings of A Royal Society Discussion Meeting held on
4 and 5 February 1987 / organized and edited by M. V. Berry, I. C. Percival, and N. O. Weiss.
p. cm.   ''First published in Proceedings of the Royal Society of London, series A, volume 413
(no. 1844) . . .''—T.p. verso. ISBN 0-691-02423-5
1. Chaotic behavior in systems—Congresses. I. Berry, Michael V. II. Percival, Ian, 1931–
III. Weiss, N. O. (Nigel Oscar). IV. Title   [Q172.5.C45R69   1987a]
003—dc19 88-32657 CIP